DARWIN, MARX, AND FREUD

Their Influence on Moral Theory

THE HASTINGS CENTER SERIES IN ETHICS

ETHICS TEACHING IN HIGHER EDUCATION
Edited by Daniel Callahan and Sissela Bok

MENTAL RETARDATION AND STERILIZATION
A Problem of Competency and Paternalism
Edited by Ruth Macklin and Willard Gaylin

THE ROOTS OF ETHICS: Science, Religion, and Values
Edited by Daniel Callahan and H. Tristram Engelhardt, Jr.

ETHICS IN HARD TIMES
Edited by Arthur L. Caplan and Daniel Callahan

VIOLENCE AND THE POLITICS OF RESEARCH
Edited by Willard Gaylin, Ruth Macklin, and Tabitha M. Powledge

WHO SPEAKS FOR THE CHILD: The Problems of Proxy Consent
Edited by Willard Gaylin and Ruth Macklin

ETHICS, THE SOCIAL SCIENCES, AND POLICY ANALYSIS
Edited by Daniel Callahan and Bruce Jennings

IN SEARCH OF EQUITY
Health Needs and the Health Care System
Edited by Ronald Bayer, Arthur L. Caplan, and Norman Daniels

DARWIN, MARX, AND FREUD: Their Influence on
Moral Theory
Edited by Arthur L. Caplan and Bruce Jennings

ABORTION: Understanding Differences
Edited by Sidney Callahan and Daniel Callahan

A Continuation Order Plan is available for this series. A continuation order will bring delivery of each new volume immediately upon publication. Volumes are billed only upon actual shipment. For further information please contact the publisher.

DARWIN, MARX, AND FREUD

Their Influence on Moral Theory

Edited by

ARTHUR L. CAPLAN
and
BRUCE JENNINGS

The Hastings Center
Institute of Society, Ethics, and the Life Sciences
Hastings-on-Hudson, New York

Plenum Press • New York and London

Library of Congress Cataloging in Publication Data

Main entry under title:

Darwin, Marx, and Freud.

(The Hastings Center series in ethics)
Includes bibliographical references and index.
1. Ethics, Modern—20th century—Addresses, essays, lectures. 2. Darwin,
Charles, 1809–1882—Ethics—Addresses, essays, lectures. 3. Marx, Karl,
1818–1883—Ethics—Addresses, essays, lectures. 4. Freud, Sigmund,
1856–1939—Influence—Addresses, essays, lectures. I. Caplan, Arthur L. II.
Jennings, Bruce, 1949–   . III. Series.
BJ319.D37   1984   170′.92′2   84-1909
ISBN 0-306-41530-5

©1984 The Hastings Center
Institute of Society, Ethics, and the Life Sciences
360 Broadway
Hastings-on-Hudson, New York 10706

Plenum Press is a division of
Plenum Publishing Corporation
233 Spring Street, New York, N.Y. 10013

Contributors

EUGENE B. BRODY is Professor of Psychiatry and Human Behavior at the University of Maryland Medical School. He is Editor-in-Chief of the *Journal of Nervous and Mental Disease* and President (1981–83) of the World Federation for Mental Health. His most recent book is *Sex, Contraception and Motherhood in Jamaica.*

ARTHUR L. CAPLAN is Associate for the Humanities at The Hastings Center. He is the editor of *The Sociobiology Debate* and *Concepts of Health and Disease* and the author of numerous articles in the philosophy of science and bioethics.

ANTONY G. N. FLEW is Professor of Philosophy at York University, Toronto, Canada and Emeritus Professor at the University of Reading, Reading, England. He is the author of *Evolutionary Ethics, A Rational Animal,* and several other books.

MICHAEL HARRINGTON is Professor of Political Science on the faculty of Queens College and CUNY Graduate Center and is the National Chair of Democratic Socialists of America. He is author of several books, including *The Other America, Socialism, Fragments of the Century,* and *Decade of Decision.*

ROBERT R. HOLT is Professor of Psychology at New York University and is the founder and former director of the NYU Research Center for Mental Health. He is the author of *Methods in Clinical Psychology* (2 vols.).

GERALD N. IZENBERG is Associate Professor of History at Washington University. He is the author of *The Crisis of Autonomy: The Existentialist Critique of Freud*.

BRUCE JENNINGS is Associate for Policy Studies at The Hastings Center. His publications include several articles on the history of political theory, and he is co-editor of *Ethics, the Social Sciences, and Policy Analysis*.

LEON R. KASS is Henry R. Luce Professor of the Liberal Arts of Human Biology at the University of Chicago. His articles have appeared in several journals, including *Science, The Public Interest, The Journal of the American Medical Association*, and *Commentary*.

ERNST MAYR is Alexander Agassiz Professor, Emeritus, at Harvard University. He is author of *Animal Species and Evolution, Evolution and the Diversity of Life*, and *The Growth of Biological Thought*.

ELIZABETH RAPAPORT is Associate Professor of Philosophy at Bennington College in Bennington, Vermont. Her publications include several articles on ethics, politics, and feminist theory.

ALLEN W. WOOD is Professor of Philosophy at Cornell University. He is author of *Kant's Moral Religion, Kant's Rational Theology*, and *Karl Marx*.

Acknowledgments

The papers in this volume are based upon a series of presentations at The Hastings Center which were underwritten by a generous grant from the Humanities Division of the Rockefeller Foundation. Mary Gualandi and Eva Mannheimer deserve special praise for their willingness to type numerous revisions of the papers included here. Also, the editors would like to thank Daniel Callahan and Willard Gaylin, respectively Director and President of The Hastings Center, for their encouragement and support of this project. Finally, we would like to thank the contributors to the volume for their patience in attending to the various editorial suggestions and requests that were made of them.

Contents

PART THREE   FREUD

Ethical Theory and Social Science

The Legacy of Darwin, Marx, and Freud

ARTHUR L. CAPLAN AND BRUCE JENNINGS

I

Charles Darwin, Karl Marx, and Sigmund Freud incontestably belong to that very small group of nineteenth- and early-twentieth-century thinkers whose work has left an indelible mark on our own thought. With an intellectual boldness and creative insight rarely matched in Western intellectual history, these men and the theoretical movements they inspired have fundamentally reoriented our modern understanding of biological nature, society, and the human mind. In so doing, they effectively rewrote the agenda of subsequent philosophy and human science.

The influence of Darwin on Marx and of both on Freud has long been a subject of considerable interest to intellectual historians, just as the effort to formulate a theoretical synthesis that would somehow fuse their respective theories into some unified or at least hyphenated whole has long provided a major growth industry for social philosophers and critical theorists. The results of these inquiries—particularly the latter—have been inconclusive at best, but the motivating impulse behind them is easy to see. If, in an age of increasingly specialized knowledge and fragmented perspectives, there is to be any hope of obtaining a comprehensive and coherent understanding of the human condition, we must somehow weave together the biological, sociological, and psychological components of human nature and experience. And this cannot be done—indeed, it is difficult to even make sense of an *attempt* to do it—without first settling our accounts with Darwin, Marx, and Freud.

The legacy of these three thinkers continues to haunt us in other ways as well. Whatever their substantive philosophical differences in other respects, Darwin, Marx, and Freud shared a common, overriding intellectual orientation: they taught us to see human things in historical, developmental terms. Philosophically, questions of being were displaced in their works by questions of becoming. Methodologically, *genesis* replaced teleological and essentialist considerations in the explanatory logic of their theories. Darwin, Marx, and Freud were, above all, theorists of conflict, dynamism, and change. They emphasized the fragility of order, and their abiding concern was always to discover and to explicate the myriad ways in which order grows out of disorder. For these reasons their theories constantly confront and challenge the cardinal tenet of our modern secular faith: the notion of progress.

To be sure, their emphasis on conflict and the flux of change within the flow of time was not unprecedented; its origins in Western thought can be traced back at least as far as Heraclitus. But, drawing imaginatively on romanticism, historicism, and other similar undercurrents of nineteenth-century thought, they did give that ancient perspective its most systematic and uncompromising modern expression. In their hands, indeed, this perspective became *the* constitutive perspective of Western modernity itself. And so it remains today. Moreover, in strikingly parallel ways, each of them self-consciously attempted to take what had traditionally been a poetic, metaphysical emphasis on becoming and to transform it into a science, in the modern sense of the term.

For Darwin, Marx, and Freud, not "objects" but "processes" became the principal units of scientific investigation. Their theories attempted rationally to comprehend and demystify the emergence of order out of disorder. And they aspired to discover the developmental "laws" governing the emergence of the functional present out of a chaotic and unstable past. In this respect, their texts provide us with rich sites, full of interpretive possibilities, where the two great currents of nineteenth-century thought—romantic historicism and positivism—encounter one another in an exceedingly complex and often volatile union.

This achievement had momentous consequences, and it is probably the single most important reason why the work of these men continues to have such a significant impact on contemporary culture. It is easy enough today to reject some of the specific conclusions they reached and answers they gave, and in this respect many aspects of their original work may be said to have been superseded. What is still much harder if not impossible to do, though, is to abandon the overall framework of analysis they established and to reject the basic questions they posed. Moreover, the significance of those questions—about biological and behavioral adaptation, class conflict, and the formative experience of the human psyche in the family and the broader culture—has not diminished during the last century but has, if anything, taken on added force.

The prospect of ecological catastrophe, revolutionary turmoil abroad and economic decline at home, and radical changes in family structure and sexual mores—all these issues and more make the basic Darwinian, Marxian, and Freudian questions more insistent and urgent than ever. And the fundamental significance of these questions, together with our growing anxiety that we cannot really understand or cope with them, makes a reexamination of the basic intellectual framework we have inherited and used for the last century something more than a merely antiquarian or methodological enterprise. It is little wonder, then, that when we return to the texts of Darwin, Marx, and Freud, we do so not only to learn something about them but also, and more importantly, to learn something about ourselves.

The purpose of this book is to examine one important but relatively neglected aspect of the enduring connection between the work of Darwin, Marx, and Freud and contemporary thought—their influence on ethical theory. Now, in one sense it is quite misleading and untrue to say that this topic has been neglected. Recent moral philosophers, particularly those working within the framework of analytic philosophy, certainly have not paid much explicit attention to Darwin, Marx, or Freud; but beyond the narrow confines of analytic ethics, a host of other thinkers have purported to draw out the moral implications of the Darwinian, Marxian, and Freudian perspectives. In short, just as there has been an enormous amount of attention paid by biologists, social scientists, and psychiatrists to the importance of Darwin, Marx, and Freud as scientists, so too there has been a huge if not always rich literature on the moral significance of their work.

Nor is it at all surprising that this should be the case. Although neither Darwin, Marx, nor Freud constructed anything resembling an ethical theory, they were all, to a greater or lesser extent, explicitly concerned with social and political issues that had profound human consequences. Like the most important ethical theorists of their time, John Stuart Mill and Henry Sedgwick, they always kept the basic questions of human happiness and well-being at the center of their attention. Moreover, at a deeper level, their theories drastically altered the traditional understanding of human nature, human action, and the relationship between the self and society—fundamental matters that have always been the basis of moral philosophy.

Nonetheless, despite the fact that a great deal has been written over the years about the moral implications of Darwin's, Marx's, and Freud's theories, and, to a much greater extent, about the moral implications of the intellectual and social movements they inspired, many questions about their significance for contemporary ethical theory remain unanswered. In the first place, as Antony Flew, Michael Harrington, and Robert Holt argue, past attempts to define the moral implications of their work have been deeply flawed by specious reasoning and incorrect textual interpretation. As often as not, subsequent polemicists and ideologues have misused the intellectual authority of Darwin, Marx, and Freud to establish normative conclusions that have no valid basis in the "sacred" texts. Much careful reasoning and exegesis is necessary to clear the air and cut through all the false starts that have accumulated over the years. The best recent scholarship on Darwin, Marx, and Freud has begun to do this, and the chapters in this volume aim to carry on that task.

Second, it is important to disentangle three related but logically separate topics: (1) Darwin's, Marx's, and Freud's own ethical and moral views; (2) the question of their influence on everyday morality; and (3) the question of their influence on subsequent ethical theorizing. The first is primarily a matter of textual exegesis. The second is basically a matter of intellectual history and the sociology of knowledge. And the third is a conceptual and philosophical issue, which raises a question about the essential compatibility between the traditional presuppositions of ethical theory and the way Darwin, Marx, and Freud would have us understand human agency, rationality, motivation, and the influence of social institutions or hereditary predispositions on human behavior. It is this third, conceptual issue that has been largely neglected. After attending to the logical confusions of past discussions and after sorting out the different kinds of questions that have too often been conflated, the discussions in this book all attempt to come to grips with this basic philosophical issue.

Before turning to the contributors' analyses, however, something more needs to be said about why a better understanding of this particular issue is important and about why the legacy of Darwin, Marx, and Freud—as it has been contained in twentieth-century social science—is especially crucial to the future of ethical theory today.

II

During the last ten years, the study of ethics has gone through a truly remarkable period of transition and renewal. After several decades of relative disinterest and neglect, ethical theory has once again become one of the most active and intellectually exciting branches of philosophical inquiry. This renewal has affected the study of ethics at all levels. At moral philosphy's most abstract and formal level of analysis, which is generally referred to as "metaethics," basic questions concerning the logic, cognitive significance, meaning, and rationality of moral discourse have all been reexamined in ways that shed new light on the theoretical foundations of ethics and its relationship to other modes of human knowledge in the arts and sciences. These recent developments in the field of metaethics have been important for ethical theory because they have persuasively called into question two earlier doctrines whose dominance in Anglo-American academic philosophy during the middle decades of the twentieth century had directed most philosophers' attention away from the traditional questions of substantive ethical theory. These doctrines were logical positivism and emotivism.

Briefly stated, logical positivism held that ethical claims are inherently subjective and devoid of cognitive significance because they cannot be logically deduced from empirically verifiable statements of fact. Emotivism held that the meaning of ethical terms is determined by the emotionally based, subjective preferences of the speaker. The upshot of both these positions was that ethical claims are not subject to rational discussion or to public, intersubjective criteria of evaluation. Recent challenges to these doctrines and alternative accounts of ethical judgment and the meaning of moral terms have provided a good deal of the philosophical warrant for the new work in ethical theory that has blossomed in recent years.

Second, at its more substantive and concrete level of analysis—"normative ethics," where particular moral principles, ideals, and virtues are formulated for the evaluation of individual conduct and social institutions—ethical theory has also opened up a whole series of new lines of investigation. The basic structures of theory formation and rational argumentation in ethics have been reexamined and several innovative theoretical positions have been developed on justice, human rights, equality, and obligation. Moreover, moral philosophers have increasingly turned to the study of traditional normative works—such as Aristotelian, natural law, Kantian, and utilitarian theories—not simply as matters of historical or antiquarian interest but rather with the intention of revitalizing these older traditions and using them to help us restore our moral bearings in a contemporary world so complex, so problematical, and so filled with destructive potential that it continuously pulls us in the direction of ethical nihilism, relativism, and despair.

Finally, impetus for the current renewal of interest in ethical theory and normative analysis has come from the rise of research and teaching in the field commonly referred to as "applied ethics," where various ethical principles and ideals are brought to bear on practical decision making and the resolution of moral dilemmas in specific institutional settings, such as governmental policymaking or the provision of professional services to clients. These studies—which typically focus on a detailed examination of particular cases and trace the institutional, social, and cultural sources of value conflicts—often provide a testing ground for the adequacy of competing normative theories. They help us to better understand what is involved in translating a general, theoretical principle into a specific course of conduct and to see what the human meaning of abstract theoretical conceptions is when those conceptions take shape in concrete social relationships.

To some extent, recent work in the first and third of these three branches of moral philosophy—metaethics and applied ethics—has provided much-needed support for normative ethics, and new work in ethical theory has begun to display both increasing methodological rigor and a more realistic awareness of the institutional structures within which the courses of action prescribed by an ethical theory can be carried out. That is to say, ethical theory no longer addresses itself solely to the character or actions of an individual moral agent, isolated from any concrete and particular social and historical context. For this reason, ethical theory is moving, albeit hesitantly, toward a common ground with political theory and the social sciences. By exploring the implications for the study of ethics contained in the work of three major thinkers who have shaped our modern social-scientific world view—Charles Darwin, Karl Marx, and Sigmund Freud—the chapters in this volume are offered as a first step toward an examination of some of the promises and pitfalls of this common ground between ethics, political theory, and social science.

As the following chapters will demonstrate clearly, the question of the relationship between ethical theory and the scientific study of social institutions and human behavior is an exceedingly knotty one, with more than its share of logically suspect connections, crude reductionism, category mistakes, and intellectual dead ends. But this is a relationship that we can no longe afford to neglect, if only because, in the years ahead, work on ethics and ethical theory promises to take on a renewed measure of importance and interest, not only in the domain of academic philosophy but also in the broader realm of social and political life. This event is not unprecedented, of course; nor, considering the times in which we live, should it be particularly surprising. On many occasions in the past, the history of Western philosophy has been punctuated by similarly creative bursts of intellectual energy directed toward ethics, often in the aftermath of rapid social or cultural change and political crisis. And surely in the current renewal of interest in ethical theory in the United States, we are witnessing a philosophical attempt to come to grips with the widespread personal, cultural, and political uncertainty of our own troubled times. As ethical theorists return to substantive normative concerns and thus join the chorus of voices engaged in public discourse on social issues, they will be given a serious hearing and perhaps even a certain measure of intellectual authority.

Indeed, this has already happened to a large extent. Many of the most central and troubling public issues of recent years— such as international human rights, sexual and racial discrimination, abortion, capital punishment, and civil disobedience— have been marked by the fact that the debates surrounding them have not simply taken place in the more or less standard ideological terms of partisan politics but have also involved explicit appeal to specific ethical theories drawn from various philosophical and theological writings. And in the area of health care and biomedical research, ethical theories have had a particularly direct impact, leading to reforms that have quite radically altered the law and informal norms concerning the definition of death, informed consent, and the treatment of research subjects. In each case, the public issues and controversies can scarcely be described, let alone understood, without a thorough familiarity with the ethical theories underlying them. These issues are not behind us, and other, similar issues, equally perplexing and equally dependent upon ethical theories for their formulation, are waiting just over the horizon.

Thus, increasingly, ethical theory will become an intellectual force to be reckoned with in the life of our society. This prospect raises, in turn, many pressing and difficult questions of its own about the adequacy of what might be called our own philosophical literacy as a democratic citizenry. How well prepared are we, for example, to answer—or even to think about—the following kinds of questions: What is the nature of an ethical theory? Are theories in ethics like theories in the sciences? How are ethical theories arrived at? Can they be "proven" or "falsified"? If so, how? If not, are there any rational evaluative criteria to use in deciding between two rival theories? If it is rational to believe in a given theory, is it possible to apply it rationally in social life? What then is the relationship between ethical theory and moral practice? And what is the relationship between an ethical theory and the prevailing moral norms and traditions of a society? Can ethical theories provide new knowledge and insight, or do they necessarily reflect and legitimate currently prevailing beliefs? And, finally, can ethical theories be understood in exclusively philosophical terms? Or are they best seen as epiphenomena governed by some more basic, underlying process such as class struggle, unconscious psychological conflict, or biological evolution?

Despite the recent surge of interest in ethical theory, the

answers to questions such as these remain elusive. The intellectual situation of moral philosophy today is paradoxical; equally important, the likely consequences of the impending political uses of ethical theory are uncertain. Intellectually, the situation of moral philosophy is one of transition and flux. Recent work in metaethics has provided an opening for the development of new normative theories and their application by overthrowing the earlier tenets of positivism and emotivism, but as yet no authoritative metaethical position has been established to take the place of these discredited views. As a result, normative ethics has flourished in a climate of opinion where everyone seems convinced that theoretical inquiry in ethics is a legitimate enterprise but no one seems certain about the precise philosophical standards that ought to govern it. Concomitantly, the political significance of work in ethical theory is clouded by the extreme ambivalence of our culture's response to ethics. This ambivalence holds the key to the future of ethical theory, and we believe that its source and nature must be clarified before the possible political significance of the revitalization of moral philosophy can be properly assessed. This ambivalence can be described in the following way.

On one hand, there is today an evident willingness in the United States and other Western societies to acknowledge the pertinence of ethical analysis to the major issues facing us and to use moral categories as the medium through which we define our common purposes and our social relationships. This attests to our culture's continuing belief in the autonomous moral agency of the human individual and in the power of moral ideas and ideals to shape and guide conduct. The notions of autonomy, rationality, and responsibility have been the cornerstones of ethical theory in the Western philosophical tradition; insofar as they can continue to provide a serviceable framework for our own self-understanding as social beings, ethical theory will remain a viable, even indispensable, component of our public discourse.

On the other hand, however, this positive reception of ethical theory is offset by an equally deep-seated and skeptical negative attitude conditioned by many formidable and long-standing intellectual movements that, in their cumulative effect, have cast radical doubt upon the adequacy of the notions of autonomy, rationality, and responsibility as serviceable categories for human self-understanding. Foremost among these movements, at least during the last hundred years, has been the rise of the modern social and behavioral sciences. Within the social sciences, some of the strongest reasons for skepticism about the adequacy and legitimacy of ethical theorizing have their roots in the intellectual legacy of Darwin, Marx, and Freud.

There are two principal reasons why the modern social sciences pose a challenge to the viability of traditional ethical theory and why, in a culture permeated by the social-scientific consciousness, the serious revival of ethical theory is bound to provoke a confrontation between seemingly antithetical perspectives. First, historically the social sciences have been deeply committed to positivism in their quest for deterministic, nomothetic explanations of human behavior and in their insistence that rational knowledge is limited to the realm of empirically verifiable, descriptive statements of fact. In line with this commitment, the social sciences have generally adopted a posture akin to ethical emotivism and have treated prescriptive values as mere expressions of psychological affect, material self-interest, or subjective preference. While, as we have indicated, positivism's and emotivism's attack on traditional moral philosophy has been countered by recent work in metaethics, a good deal of the spirit and substance of that attack continues to flourish in the social sciences, and their influence continues to foster a deep division between descriptive and evaluative discourse in our culture.

If we are to reassess this division and overcome some of the confusion it engenders in contemporary social science and ethical theory alike, it would be useful to explore the roots of positivism and emotivism in the work of Darwin, Marx, and Freud and in the traditions of inquiry they founded. In raising this point we do not mean to suggest that either Darwin, Marx, or Freud propounded positivism or emotivism as systematic doctrines. But their work—and their legacy—does mount a skeptical attack on the purported rationality and objectivity of moral reasoning and ethical judgment. And at least Marx and Freud, if not Darwin, launched a self-conscious attempt to use descriptive science to unmask seemingly universalistic normative doctrines which, they believed, really concealed particularistic claims of individual or class interest. Since they did not attempt to construct a universalistic ethic that could withstand this sociological or psychological unmasking, it is not difficult to see why, in later years, their work has been seen as antithetical to ethical theory or moral philosophy in the traditional sense.

At the same time, however, Marx and Freud were both, in many respects, "moralists" themselves. The basic theoretical categories they employed respected no firm boundary line between scientific description and moral evaluation. A close study of their work with this problem in mind can perhaps give us fresh insight into the ambiguity of our own present understanding of the relationship between descriptive and evaluative discourse—social science and ethical theory.

The second and in some ways more interesting and crucial reason for the essential tension between ethical theory and the social sciences has to do with the basic picture presented by the social sciences of the human individual in modern society. Here the point at issue involves not matters of methodology and epistemology but the meaning and pertinence of those traditional conceptions of autonomous, rational, and responsible agency upon which ethical theory has been based. By and large, the social sciences have given us a powerful and systematic account of the social, psychological, and even biological facets of the human condition in which the categories of autonomy, rationality, and responsibility have been replaced by a set of categories designed to accentuate the external, material, or irrational determinants of human behavior. Developed in response to the massive social changes provoked by nineteenth-century capitalism and the rise of the bureaucratic nation-state, the social sciences have radically transformed the Western image of humanity and have greatly increased our awareness of the ways in which individuals are subject to the play of forces that are largely beyond their understanding or control.

In short, the social sciences have tended to shatter the conceptual framework of ethical theory, either by reinterpreting its basic categories so that their meaning is altered beyond recognition or else by dispensing with those categories altogether. In the world described by the social sciences, the traditional notion of autonomous moral agency is at best exceedingly problematic and at worst incoherent. And yet it is precisely this traditional notion that seems to be coming back into its own with the current renaissance of ethical theory and applied ethics.

If contemporary ethical theory is to retain its hard-won philosophical legitimacy and respectability and if it is to succeed in its efforts to emerge from the confines of the academy and play a more direct role in the public discourse of our society, moral philosophers and social scientists alike will have to reexamine the roots of the tension between their disciplines and, through a better historical and conceptual understanding of that tension, seek ways to resolve it. One way to begin to do that is to return to the origins of the modern social-scientific outlook and see how the ethical paradigm of human action and the social scientific paradigm began to diverge. Such a quest must begin with a careful examination of the moral and intellectual legacy of Darwin, Marx, and Freud.

III

The first trio of papers in this volume, by Professors Flew, Mayr, and Kass, address themselves to the impact of Darwinism on morality. None of the authors is particularly impressed with the influence of evolution on Darwin's own moral viewpoints. Professor Flew notes that to some extent progressivism and perfectionism occasionally influence Darwin's thoughts

about the implications of his theory; but, for the most part, all
the authors agree that the major impact of Darwinism on ethics
had to do with its impact on others. In particular, Professor
Flew draws our attention to the impact of the Darwinian world
view upon both the religious thought and political theorizing
of the nineteenth century. He believes that Darwinism basically undercuts the foundations of a theology based upon literal
accounts of the Bible and the unique and special status of
humankind. Professor Kass takes issue with this position, arguing that, regardless of its historical impact, Darwinism does
not necessarily contradict a naturalistic outlook on the purposes and goals that humanity should strive to achieve. He
argues that only the crudest and, in a peculiar way, most
scientistic view of religion is placed in jeopardy by a Darwinian
world view. Professor Flew also is concerned to show the del-
eterious impact Darwinism had on certain forms of nineteenth-
and early-twentieth-century moralizing, particularly that of social Darwinism. Professor Mayr agrees that to some extent the
Darwinian impact on everyday moral thinking during this period was unfortunate, but both Flew and Mayr argue that
Darwinism, properly understood, can be a liberating source of
inspiration for contemporary ethics. Professor Flew bemoans
the fact that for much of the twentieth century those involved
in academic ethics have remained immune to the implications
of a Darwinian world view. Again, Professor Kass worries that
an attempt to secularize ethics by means of a Darwinian outlook may ultimately fail to recognize the teleological dimen-
sions essential to the moral point of view.

The next three contributors to this volume—Michael Harrington, Elizabeth Rapaport, and Allen Wood—examine Marx
and his moral views as well as the impact of Marxism upon
contemporary ethical theorizing. Professor Harrington is concerned to show that while Marx, like Darwin, spends very little
time moralizing in his writings, he nevertheless had a powerful
commitment to key moral values throughout his work. Thus,
Harrington does not view Marx as an opponent of ethical theory *per se,* nor does he equate Marx's ethics with worker-

consciousness. Rather, he argues that freedom, self-control, liberation, and emancipation motivate much of Marx's thought, and he takes these notions to be the central and distinctively moral values in Marx's work.

Allen Wood takes issues with this view of Marx's ethics. He argues that while it may be true in a very general sense that Marx has a commitment to certain values or goods, such as liberation or freedom, it makes no sense to say that Marx recognized the need for morality simply because he manifests a commitment to certain central values. Wood sees Marx as being committed to the cultivation of social movements, not moral ideals. Moreover, he sees Marx as being entirely hostile to systems of ethical theory and to the necessity of motivating action by means of moral norms. Wood thinks that the only way Harrington is capable of preserving morality in Marx's writings is by inflating the notion of morality to the point where it includes almost any valuable or good thing. And this Wood simply rejects as an appropriate definition of either ethics or morality.

Harrington is concerned to point out the unhappy impact of Marxism upon the ethical debates that followed among Stalinists and others in the twentieth century. Harrington notes that, particularly in the Soviet Union, Communist rulers were able to use crude interpretations of Marxist ethics to justify totalitarian and autocratic governance of the population. Harrington notes that in Western Marxist circles, debates about morality quickly became enmeshed in arguments about human nature; in this way there would appear to be some overlap to the impact of Marxism upon its followers and the impact of Darwinism upon its contemporary devotees.

Elizabeth Rapaport notes that one of the major values of Marxism for contemporary moral philosophy is as a tool for understanding the lack of value consensus that seems to be present in many societies. Historical materialism is helpful, she argues, in revealing the sociological and economic foundations for much of the value discontent that appears in American society today. Harrington agrees and argues that in some ways Marxism—and the dialectical reasoning it embodies—might prove to be a useful antidote to the kind of crude materialism and hedonism that seems to govern so much of ordinary moral practice. Rapaport takes issue with Harrington about the possibility of using Marxism as a liberating philosophy as against hedonism, since she feels that the failure of Marxism as a social science undercuts its ability to provide a moral foundation for the sort of democratic socialism which Harrington hopes to derive from the legacy of nineteenth-century Marxist analysis.

It is interesting to note that the scholars writing on Darwin and those attempting to analyze Marx agree that the view of human nature provided by both of these thinkers holds great promise for enlivening contemporary moral philosophy. Moreover, both groups of scholars seem to agree that the influence of Marx and Darwin was far greater in terms of modifying our self-conception as historical creatures whose morality must be to some extent contingent upon history than it was in terms of any specific normative or prescriptive theories that either of these thinkers advanced.

The last three contributors to the book—Robert Holt, Eugene Brody, and Gerald Izenberg—are in rather close agreement about their analysis of the impact of Freud and Freudianism on ethics. Professor Holt argues that Freud, like Marx, was concerned to use his theories to help liberate individuals from unsuspected and unseen restraining forces. He argues that while Freud was not hostile to all forms of morality, he nevertheless was quite critical of certain religious beliefs and moral taboos. Unlike Darwin and Marx, Freud took one of his primary tasks to be the explanation of morality. Whereas Darwin viewed ethics as an autonomous enterprise outside the realm of biological constraints and Marx found an analysis of morality more or less peripheral to his interests, Freud saw the analysis of moral beliefs, practices, and customs as central to the task of psychoanalytic theory. Freud explained morality as an effort to control certain unruly wishes and desires and thought that through psychoanalysis individuals could be allowed to become aware of these wishes and desires and thereby exercise more choice with regard to their life-style and behaviors.

Holt also emphasizes that Freud was quite committed to a rather conservative ethics in the practice of psychoanalysis. While many of Freud's followers, as Izenberg also notes, saw Freudian theory as licensing the complete abandonment of prescriptive morality, particularly with regard to sexual behavior, Holt notes that Freud felt that such a suspension of moral judgment was only appropriate within the psychoanalytic setting. Holt also comments that Freudianism—with its powerful commitment to individualism, emancipation, and liberation—was very much a product of the late nineteenth century, and that these values were also prominent in Marx's theorizing.

Eugene Brody sees a powerful legacy of Freudianism in contemporary psychoanalysis, with its emphasis on the therapeutic value of requiring patients to take responsibility for their behavior. Brody argues that to some extent self-knowledge can lead to a greater assumption of personal responsibility for one's activities and actions, thereby creating and legitimating a moral sense within the patient. Gerald Izenberg worries, however, that the emphasis on responsibility in contemporary psychoanalysis is perhaps too reinforcing of a kind of narcissism that extols, or even unrealistically distorts, the power of the individual to struggle against or overcome social conditions and factors. Professor Holt argues that Freudianism can be used to formulate a more dynamic world view in which materialism can be supplanted by a holistic humanism; this is in many ways similar to the views expressed by Flew and Mayr about the power of Darwinism to create a liberating world outlook. It is interesting to note that Holt, Flew, and Harrington all agree that the major legacy of Darwin, Marx, and Freud to contemporary ethics is in the power of their theories to amplify human choice through the maximization of human understanding.

PART ONE

Darwin

CHAPTER ONE

The Philosophical Implications of Darwinism

ANTONY G. N. FLEW

> *The Origin of Species.* . . . With the one exception of Newton's *Principia* no single book of empirical science has ever been of more importance to philosophy than this work of Darwin.
>
> Josiah Royce (*The Spirit of Modern Philosophy*, p. 286)

> The Darwinian theory has no more to do with philosophy than has any other hypothesis of natural science.
>
> Ludwig Wittgenstein (*Tractatus Logico-Philosophicus*, § 4, p. 1122)

When the word "philosophy" in each of these two apparently contradictory sentences is given the appropriate sense, both express plain and entirely compatible truths, truths that are both, in their different contexts, important. In the first, the relevant sense is wide and untechnical. It is in this original and most common understanding that biographers devote chapters to the philosophy of their subjects, professional associations invite leading figures to address ceremonial occasions on their personal philosophy of whatever it may be, and editors of serious general journals commission contributions to symposiums on the philosophical implications of new theoretical developments. In the second sentence quoted, the relevant sense is narrow and technical. In this specialist sense we could with equal truth say (1) that while most of *The Laws* and much of *The Republic* is not philosophy, *Theaetetus* is

almost nothing but, and (2) that of the comparatively few pages
of philosophy in Hume's second *Inquiry*, most are of intent
relegated to the appendixes.[1]

The two understandings of philosophy thus illustrated
should both be acceptable. Although they are very different,
the second does have implications for the first. Here the most
obvious example is discussion of the logical relations, or lack
of relations, between "ought" and "is." This is a paradigm case
of philosophy in the second interpretation. Yet any answer
given must be crucially relevant to questions about the philosophical implications of Darwinism—issues that are just as
paradigmatic of the first interpretation. So all the present paper
will be philosophical in one or the other understanding and
much of it in both.

I. THE ESSENTIALS OF DARWIN'S THEORY

Charles Darwin (1809–1882) published his first book in
1859. It is from this epoch-making and epoch-marking land-
mark that all our discussions have to begin. Yet even Darwin's
apt and carefully chosen main title is rarely given in full: *The
Origin of Species by Means of Natural Selection*. There was
also a subtitle, which has since, for all who were alive and
alert in the late thirties and early forties, acquired a sinister
ring: *Or the Preservation of Favoured Races in the Struggle
for Life*. Darwin's theory is evolutionary inasmuch as it asserts,
and provides an account of, the smoothly developmental origin
of species. Evolution is here to be contrasted with the previously dominant notion of the fixity of species, all of which had
presumably been specially created in substantially their present form. That is assumed, for instance, in the picturesque
creation stories of Genesis, stories illustrated by—among so
many others—William Blake. *Natural selection* is the key term
in Darwin's account of how evolution, in this sense, has occurred and still is occurring.

[1]A. G. N. Flew, *Philosophy: An Introduction* (Buffalo, N.Y.: Prometheus,
  1979), chap. I.

None of the various notions incorporated in Darwin's conceptual scheme was by itself new. The originality and the greatness of his achievement lies elsewhere. He was not—nor, of course, did he ever claim to be—the first to assert the evolution of species as opposed to their special creation: "The general hypothesis of the derivation of all present species from a small number, or perhaps a single pair, of original ancestors was propounded by the President of the Berlin Academy of Sciences, Maupertuis, in 1745 and 1751, and by the principal editor of the *Encyclopédie*, Diderot, in 1749 and 1754."[2]

Nor was Darwin the first to introduce into a biological context the ideas either of natural selection or of a struggle for existence. Both can be found in the Roman poet Lucretius during the first century B.C., in an account of how, in the beginning, our mother earth produced both all the kinds of living things that we now know and many other sorts of ill-starred monstrosity. But with these latter "it was all in vain . . . they could not attain the desired flower of age nor find food nor join by the ways of Venus." The poet concludes: "And many species of animals must have perished at that time, unable by procreation to forge out the chain of posterity; for whatever you see feeding on the breath of life, either cunning or courage or at least quickness must have kept . . . from its earliest existence" (Lucretius, V, 845–8 and 855–9). Lucretius, too, was a disciple, clothing in Latin verse ideas he had himself learned from the fourth-century Greek Epicurus, who was here in his turn drawing on such fifth-century sources as Empedocles of Acragas.[3]

Darwin himself was well aware of, and indeed took delight in pointing to, many of these partial anticipations—especially, perhaps, those compassed by his own engaging grandfather Erasmus. But attention to such anticipations must not be al-

---

[2] A. O. Lovejoy, *The Great Chain of Being* (New York: Harper, 1936), p. 268. Compare Lovejoy, "Some Eighteenth Century Evolutionists" in *Popular Science Monthly*, 1904.

[3] G. S. Kirk and J. E. Raven, *The Pre-Socratic Philosophers* (Cambridge, England: Cambridge University Press, 1957), pp. 336–40.

lowed to distract us from the reasons why, in his new employment, separately ancient ideas had, and deserved to have, so enormous an impact. These reasons are, first, that Darwin put it all together, assembling the scattered conceptual essentials into a single deductive scheme; and, second, that under the control of that scheme Darwin deployed an immense mass of supporting evidence, much of this the product of his own fieldwork in the years on the *Beagle* and after. In a characteristic passage of the *Autobiography*, modest yet observant, he says, " *The Origin of Species* is one long argument from the beginning to the end."[4] He had said the same in the final chapter of the book itself: "this whole volume is one long argument."[5]

It was precisely this overwhelming argument which slowly convinced the contemporary scientific world that, whatever other factors in the process might remain still to be discovered, evolution by natural selection must have occurred and be still occurring. Furthermore, overwhelming though it is, *The Origin*—quite unlike *Principia*—can be understood immediately by the intelligent but probably unmathematical layperson. The combined effect was and remains to make it imperative for anyone with any pretensions toward a rational and comprehensive world outlook to try to come to terms with Darwinism.

So let us refresh our memories. The "Introduction" indicates both that the book has a deductive skeleton and what that skeleton is:

As many more individuals of each species are born than can possibly survive; and as, *consequently* there is a frequently recurring struggle for *existence*; *it follows that* any being, if it vary however slightly in any manner profitable to itself, under the complex and sometimes varying conditions of life will have a better chance of surviving and thus be naturally selected. From the strong principle of inheritance, any selected variety will tend to propagate its new and modified form.

[4] *The Autobiography of Charles Darwin*, ed. Nora Darwin (London: Collins, 1958), p. 140.
[5] *The Origin of Species* [1859], ed. J. W. Burrow (Harmondsworth, England: Penguin Books, 1968), p. 435.

Darwin also promises that in the chapter "Struggle for Existence" he will treat this struggle "amongst all organic beings throughout the world, which inevitably follows from the high geometrical ratio of their increase."[6]

In that chapter the argument is developed:

A struggle for existence *inevitably follows* from the high rate at which all organic beings tend to increase . . . as more individuals are produced than can possibly survive, *there must in every case be* a struggle for existence, either one individual with another of the same species, or with the individuals of distinct species, or with the physical conditions of life. It is the doctrine of Malthus applied with manifold force to the whole animal and vegetable kingdom; for in this case there can be no artificial increase of food, and no prudential restraint from marriage. Although some species may be now increasing, more or less rapidly, all *cannot* do so, for the world would not hold them.[7]

Just as the struggle for existence is derived as a consequence of the combination of a geometrical ratio of increase with finite resources for living, so, in the chapter "Natural Selection," this in turn is derived as a consequence of the combination of the struggle for existence with variation. Darwin summarizes his argument here:

If . . . organic beings vary at all in the several parts of their organization, and I think this cannot be disputed; then . . . I think it would be a most extraordinary fact if no variation had ever occurred useful to each being's own welfare, in the same manner as so many variations have occurred useful to man. But if variations useful to any organic being do occur, assuredly individuals thus characterized will have the best chance of being preserved in the struggle for life; and from the strong principle of inheritance they will tend to produce offspring similarly characterized.

However, Darwin goes on, "Whether natural selection has really thus acted in nature . . . must be judged of by the general

[6] Ibid., p. 68.
[7] Ibid., pp. 116–7; italics supplied.

tenour and balance of evidence given in the following chapters." Nevertheless

> we already see how it entails extinction; and how largely extinction
> has acted in the world's history, geology plainly declares. Natural
> selection, also, leads to divergence of character; for more living beings
> can be supported on the same area the more they diverge in structure,
> habits, and constitution; of which we see proof by looking at the
> inhabitants of any small spot.[8]

II. THE CHALLENGE TO RELIGIOUS ASSUMPTIONS

In 1859 the first extrabiological implications to be per-
ceived were those having some obvious bearing upon accepted
religious ideas. Two sensitive spots were seen to be threatened.
First, it is obviously impossible to square any evolutionary account of the origin of species with a substantially literal reading
of the first chapters of Genesis. Second, a theory which promises to show that and how even the most sophisticated organs
and organisms might have developed without benefit of any
foresight or planning menaces the most persuasive pillar of
natural theology—the argument to design. Both of these threats
bear also upon morals: the first through its implications for
estimates of the nature of man; the second because it suggests
that ethics may not need in any way to take account of a
Creator.

Man No Longer a Religious Animal?

People sometimes wonder why the conflict between evolutionary biology and biblical literalism caused so much more
storm and stress than the earlier conflict between Genesis and
the evolutionary geology of Lyell; and why this second conflict,
fiercer and noisier, erupted years after the other had seemingly
subsided. The answer can be extracted by attending to the
form of Bishop Wilberforce's onslaught on Thomas Henry

[8] Ibid., pp. 169–70.

Huxley at the British Association meeting of 1860 in Oxford. Wilberforce, it will be remembered, asked whether it was on his mother's or his father's side that Huxley claimed descent from a monkey.

What had Wilberforce worried—and in this anxiety he neither was nor should have been alone—was the evolutionary threat to all traditional assumptions about the divinely established special status of our own most favored species. The details of a literal as opposed to a nonliteral reading of Genesis are perhaps of only ephemeral interest. What matters in the longer term is that the entire Bible—as well as the authorities of some though certainly not all other great religions—takes it for granted that *Homo sapiens* is the special concern of the Creator. Ours is thus somehow a species set apart, by divine authority separated both from the brutes and from the rest of animate and inanimate nature.

The Origin of Species had in fact almost nothing direct to say about *The Descent of Man*. Yet Wilberforce was right to detect the doctrine of the later book implicit in the earlier. For there in the final paragraph Darwin wrote:

Man may be excused for feeling some pride at having risen . . . to the very summit of the organic scale. . . . We must, however, acknowledge . . . *that man with all his noble qualities* . . . with his god-like intellect which has penetrated into the movements and constitution of the solar system. . . . Man *still bears in his bodily frame the indelible stamp of his lowly origin.*[9]

One consequence is that the comfortable old Cartesian compromise now falls under suspicion, and with it the Roman Catholic teaching that human souls—the essential elements distinguishing men from the brutes—are specially created. Suppose that we allow Darwin's theory to become established and to develop. Then the time will soon come when the biologists start to deny the Cartesian ghost and to incorporate

[9]*The Descent of Man* [1871] vol. I (London: Murray, 1871), p. 405; italics supplied.

its supposed powers and functions in their studies of the Cartesian machine. Putting the same point in a less flowery way,
they will not be content to abide by the ruling issued in 1953
by Pope Pius XII in the encyclical *Humani Generis*: that "the
teaching of the church leaves the doctrine of evolution an open
question, as long as it confines its speculations to the development, from other living matter already in existence, of the
human body." Nevertheless, the same pope continues, "That
souls are immediately created by God is a view which the
Catholic faith imposes on us."[10]

If these souls are allowed to be undetectable by the vulgar
methods of empirical inquiry, then no doubt such compromises
are still possible in the weak sense that the actual findings of
evolutionary biology can be made formally consistent with Roman Catholic religious requirements. No doubt too it will remain equally possible, first, to reject that particular and peculiar Roman Catholic doctrine and then to accept all those
findings while still maintaining, with strict consistency and
coherence, that evolution was God's own way of producing a
species intended to have a special status in creation. Darwinism was and remains a threat to such traditional religions: not
because it may formally falsify the old assumption of special
status; but because it makes it plumb unbelievable.

On the other hand, it is as wrong as it is common to argue
that to take a Darwinian view of *The Descent of Man* is to be
committed to saying that there cannot be any important differences between (most) people and (most of) the brutes. For
to say that this evolved or developed from that, since it presupposes that this is *not* identical with that, is precisely *not*
to say that, *really* or *ultimately*, this must be that. It is no kind
of evolutionary insight to maintain that oaks are, *in the last
analysis*, acorns.

Such reductionist misinterpretation can be found in serious scholarship as well as in works of catch-guinea popu-

[10] H. Denzinger, ed., *Enchiridion symbolorum,* 29th rev. ed. (Freiburg-in-
Breisgau, Germany: 1953), sec. 3027.

larization. It was, for instance, a lapse when the distinguished authoress of a study entitled *Darwin and the Darwinian Revolution* wrote, of his suggestions in the *Descent* that "as he earlier reduced language to the grunts and growls of a dog, he now contrived to reduce religion to the lick of the dog's tongue and the wagging of his tail."[11]

But in such best-sellers as *The Naked Ape* and *The Human Zoo*, similar reductions of the civilized to the savage and the human to the subhuman are not aberrations but the chief stock in trade. Morris, the popularizer, steps forward to reveal what his colleagues have supposedly discovered: man today is, when the chips are down, "a primitive tribal hunter, masquerading as a civilized, super-tribal citizen."[12] Generally the ape instincts overshadow everything learned. Thus, in listing conditions of intergroup violence among people, Morris explains that he has "deliberately omitted . . . the development of different ideologies. As a zoologist, viewing man as an animal, I find it hard to take such differences seriously."[13]

Not only is it both absurd and unevolutionary to present evolved products as themselves being no more or other than what they evolved from, but it is also preposterous to offer as a trophy of biological enlightenment a systematic refusal to attend to the respects in which our species differs from all the rest. To anyone thinking zoologically, surely the first three peculiarities to leap to mind must be the far-extended period between birth and maturity, the incomparable capacity for learning, and the prominence of learned as opposed to instinctual behavior. This unparalleled capacity for learning, together with its main instrument and expression, language, provides our species with a serviceable substitute for the inheritance of acquired characteristics. (I must say here in my text, and not merely in an unread note, that those last thoughts are

[11] G. Himmelfarb, *Darwin and the Darwinian Revolution* (London: Chatto and Windus, 1959), p. 307.

[12] D. Morris, *The Human Zoo* (London: Cape, 1970), p. 248.

[13] Ibid., pp. 47–8; compare the discussion in A. G. N. Flew, *A Rational Animal* (Oxford, England: Clarendon, 1978), pp. 29–33.

borrowed from Julian Huxley's *Essays of a Biologist*, which is as much a model of how the subject ought to be presented to the laity as the works of Desmond Morris are object lessons of the opposite.)

Man No Longer Unique?

The first of the two Darwinian challenges to religious assumptions has, as we have seen, a continuing purely secular interest also. For it might be thought, and in fact very often is thought, that to accept the evolutionary continuity between men and brutes is to prejudice the possibility of insisting that our species is in several important respects unique. Much indeed of the brouhaha stirred by the recent publication of *Sociobiology: The New Synthesis* was surely an expression of the fear that Edward Wilson and his colleagues must deny or at least depreciate all human uniqueness.[14] But the second of these challenges appears to have no such direct relevance to anything of continuing purely secular interest. Its relevance to our own chief present concern, morals, is indirect; insofar as Darwinism undermines the most popularly persuasive argument for the existence of God, it must make for the complete secularization of ethics.

The classical statement of that most ancient argument is found, as everybody knows, in the *Natural Theology* of William Paley. If from the observation of a watch we may infer the existence of a watchmaker, then, surely, by parity of reasoning, from the observation of mechanisms so marvelous as the hu-

[14] See, for instance, Arthur Caplan's review of the reviews published in the *Hastings Center Report* for April 1976 under the title "Ethics, Evolution, and the Milk for Human Kindness." The most furious reaction came from a collective of Boston correspondents. They were radicals upset by the challenge to their favorite, false assumptions. Nowadays such persons insist that we are all born as near as makes little matter identical—that we are almost completely creatures of and for environmental manipulation. For some discussion of these assumptions in contemporary social science, see A. G. N. Flew, *Sociology, Equality, and Education* (New York: Barnes and Noble, 1976), chaps. 4 and 5.

man eye we must infer the existence of a Great Designer, God? Paley specifically repudiates as an alternative any suggestion that

the eye, the animal to which it belongs, every other animal, every plant . . . are only so many out of the possible varieties and combinations of being which the lapse of infinite ages has brought into existence: that the present world is the relic of that variety; millions of other bodily forms and other species having perished, being by the defect of their constitution incapable of preservation. . . . Now there is no foundation whatever for this conjecture in anything which we observe in the works of nature.[15]

This favorite version of the argument to design was first formidably criticized by Hume, though Paley shows little sign of trying to come to terms with that criticism.[16] But it is Darwin and his successors who provide the decisive reinforcements. If it is possible on these lines to provide a naturalistic account of the origins of all the species of living things, then there is no other and greater kind of marvel in the universe for which we can be forced to postulate supernatural design. (It is not without reason that everywhere hard-line old-time Christians see evolutionary biology as the most dangerous science, or that in Britain members of our fading Rationalist Press Association have earned the nickname "Darwin's Witnesses.")

There is, however, another version of the argument to design that cannot be overthrown by present or promised scientific advances. While the popular version starts from particular marvels within the universe—marvels that are supposed to be naturalistically inexplicable—this version is prepared to accept all such naturalistic explanations and to take off from what might thereby be revealed of the general regularities of the universe as a whole. This alternative argument can per-

---

[15] W. Paley, "Natural Theology," in *Works,* vol. I (London: Longmans, 1938), p. 32.
[16] D. Hume, *An Inquiry Concerning Human Understanding* [1748], ed. L. A. Selby-Bigge, rev. P. H. Hidditch (Oxford, England: Clarendon, 1975), sec. XI, and *Dialogues Concerning Natural Religion* [1779] ed. N. Kemp Smith (Edinburgh, Scotland: Nelson, 1947), passim.

haps, without exercising too much sympathetic imagination, be discerned in "Way Five" of Aquinas. Its nerve is the contention that these regularities cannot be intrinsic but must instead be imposed on the universe by an extraneous Orderer, "which all men call God." The crushing reply was provided again by Hume. This is an expression of what—following Pierre Bayle—he liked to call not a Stratonian but "Stratonician atheism." It is best put as a question: "Whatever warrant could we have that the order which we discover in the universe—which is necessarily the only one we either do or could know—is not, as it appears to be, intrinsic but imposed?"[17]

III. A GUARANTEE OF PROGRESS?

The upshot of section II, therefore, is that Darwinism must tend to discomfit adherents of Platonic–Cartesian pictures of the nature of man and to reinforce the convictions of those inclined towards a Stratonician atheism. For us this points to a monistic and mortalist (and to that extent Aristotelian) view of our nature as well as to an ethics which—like that of Hume—is through and through secular, this worldly, and man-centered. In this section I shall consider the common claim that Darwinism also entails a belief in the inevitable moral progress of mankind. This claim, however, is groundless. To appreciate this and why it is so and, more important, why it could not but be so, we need to examine some of the key concepts.

Is It Really the Fittest that Survive?

The first suggestion that Darwin's theory promises progress lies in his employment of the catchy phrase "the survival of the fittest." Yet, as has by now so frequently been pointed

[17] D. Hume, *Dialogues Concerning Natural Religion*, passim. Compare A. G. N. Flew, *God and Philosophy* (London: Hutchinson, 1966), secs. 3.1–3.30, or A. G. N. Flew, "Introduction," *Malthus on Population*, vol. VI (Harmondsworth, England: Penguin Books, 1970), sec. 6.

out, the survival of the fittest is here guaranteed only and precisely insofar as actual survival is the criterion of fitness to survive. In effect, in this context, "fitness" is defined as "having whatever it may as a matter of fact take to survive." It is notorious that by other and independent standards such biological fitness may be most unadmirable. An individual or a species can have many splendid physical or mental endowments without these being or ensuring what is in fact needed for survival. Men who are in every way wretched creatures may, and all too often do, kill superb animals, while genius has frequently been laid low by the activities of unicellular beings having no wits at all.

It is in this light that we have to discount one or two overoptimistic misstatements made by Darwin himself of the implications of the theory. The chapter on instincts, for instance, ends with the sentence:

finally . . . it is far more satisfactory to look at such instincts as the young cuckoo ejecting its foster-brothers, ants making slaves, the larvae of ichneumonidae feeding within the live bodies of caterpillars, not as specially endowed or created instincts, but as a small consequence of *one general law leading to the advancement of all organic beings, namely, multiply, vary, let the strongest live and the weakest die.*[18]

In the present section so far my commentary has been all trite stuff, albeit not by the same token trivial. What is not so hackneyed is the recognition that these observations do not imply that assertions about the survival of the fittest must be empty tautologies, saying nothing about "matters of fact and real existence." To assert, for instance, that in some particular struggle for existence the fittest will survive is certainly not to promise that the outcome will be for the best. Nor is it to pick out and name the actual winners. Yet neither is it just emptily to state that it will be the winners who will win. Instead, it is to say that in the upcoming struggle for existence whoever in

---

[18] Darwin, *The Origin of Species,* p. 236; italics supplied.

fact wins will win because they happen to possess what in that particular environment turns out to be an edge of advantage over the going competition. This is, of course, a very general statement. Yet it is clearly contingent, not tautological. One thing it precludes is the conceivable, false alternative that both those who die before reproducing and those who survive to reproduce, survive or die at random.

Are the Later Always and Necessarily the Higher?

Again, traditional contrasts between higher and lower animals, combined with the more recent recognition that the former are all among the later products of the evolutionary process, may raise hopes of finding in living nature a passable substitute for Matthew Arnold's God: "Something, not ourselves, which makes for righteousness." In his famous early essay entitled "Progress, Biological and Other," Julian Huxley, launching what became a lifelong quest for such a substitute, contrived to detect several trends that he felt able to commend as progressive in whatever of the evolutionary story had by then become available. The same essay took as one of its mottoes the final sentence in the penultimate paragraph of *The Origin of Species*

As all the living forms of life are the lineal descendants of those which lived long before the Cambrian epoch, we may feel certain that . . . no cataclysm has desolated the whole world. Hence we may look with some confidence to a secure future of great length. *And as natural selection works solely by and for the good of each being, all corporeal and mental endowments will tend to progress towards perfection.*[19]

To serve Huxley's purpose, trends are no use at all. For trends can at any time cease or reverse. What he needs is at least a tendency of the kind likely to be realized, but preferably

[19] Ibid., p. 459; italics supplied.

a historicist law of development. One thinks here of the double and doubly false claim made by Engels at the grave of Marx: "Just as Darwin discovered the law of development of organic nature, so Marx discovered the law of development of human history."

Nor is the statement italicized above deducible from Darwin's theory: it follows that "endowments will tend to progress towards perfection" only when they are found in organisms not lacking something else needed to stay in the race; and, unless it is good for an individual or a species to die if it has just not got what it takes to survive, "natural selection" does not work "solely by and for the good of each being."

Notwithstanding that he had once pinned into his copy of the *Vestiges of the Natural History of Creation* a memorandum slip to warn himself "Never use the words *higher* and *lower*," Darwin still writes in his final peroratory paragraph that "from the war of nature, from famine and death, the most exalted object which we are capable of conceiving, namely the production of the higher animals, directly follows. There is grandeur in this view of life."[20]

There is indeed "a grandeur in this view of life," and it is something that should be part of the world outlook of every modern person. But what is not to be found in Darwin's theory is the kind of guarantee of progress of which he himself gave very occasional hints and that Julian Huxley and others have sought so long and so pathetically. Huxley craved, as he wrote in that first book of essays, "to discover something, some being or power, some force or tendency . . . moulding the destinies of the world—something not himself, greater than himself, with which [he could] harmonize his nature . . . repose his doubts . . . achieve confidence and hope."[21]

---

[20] F. Darwin and A. C. Seward, *More Letters of Charles Darwin*, vol. I (London: Murray, 1903), p. 114n; and *Origin of Species*, p. 459.

[21] J. Huxley, *Essays of a Biologist* [1923] (Harmondsworth, England: Penguin Books, 1939), p. 17; and compare A. G. N. Flew, *Evolutionary Ethics* (London: Macmillan, 1967), sec. III.

Human Choice Forbids All Non-Human Guarantees

The most fundamental reason why no such guarantee can be found in this or any other theory of subhuman evolution emerges as we ponder: first, the significance of the reference to Malthus in a passage quoted at the conclusion of section I; and, second, the significance too of the fact that what Darwin articulates is a theory not of selection but of natural selection. The three crucial points here are: (1) this theory, for the best of reasons, leaves out man; (2) man is the creature that can (and cannot but) make choices; and (3) it is upon the senses of these choices that the future now largely depends—not only for mankind, but also for almost all other species. Therefore, if there is any legitimate satisfaction anywhere for the secular religious cravings of Huxley and his like, it can only be in the distinctively human sciences.

The truth of the third of these points comes out most harshly when we notice why we cannot echo Darwin's conclusion that "we may look with some confidence to a secure future of great length." Armageddon apart, however, Quinton was stating the simple truth when he said: "[Man] has certainly won the contest between animal species in that it is only on his sufferance that any other species exist at all, amongst species large enough to be seen at any rate."[22]

The most illuminating way of bringing out the importance of the first two points is by approaching, as did Darwin himself, through Malthus. My treatment here will simply assert the essentials of an interpretation developed and defended elsewhere.[23] The theory of human population that Malthus published in the *First Essay* of 1798 was appropriate to natural science as opposed to human; he seems, in fact, to have held classical mechanics before his eyes as a model.

---

[22] A. M. Quinton, "Ethics and the Theory of Evolution," *Biology and Personality,* ed. I. T. Ramsey (Oxford: Blackwell, 1966), p. 120.
[23] A. G. N. Flew, *A Rational Animal,* and "Introduction," *Malthus on Population.*

His fundamental principle of population, "a prodigious power of increase in plants and animals"—the animal power to multiply "in geometrical progression"— is, therefore, construed as being not only in plants and in the brutes but also in people a power that must, in fact, inevitably be fulfilled save insofar as such fulfillment is made physically impossible by the operation of countervailing forces. No wonder that the conclusions of this *First Essay* are harsh, describing always what supposedly has happened, happens, and will happen, necessarily, ineluctably, and unavoidably. But in the *Second Essay*, which publishers and librarians misleadingly count as simply a second edition of the earlier work, Malthus introduced the notion of "moral restraint," and so "endeavoured to soften some of the harshest conclusions of the *First Essay*." Although Malthus neither spelled out nor even saw the full consequences, this ostensibly modest amendment totally transformed his theory. For the amendment makes that theory recognize what everyone with pretensions to constitute human science must recognize: the reality of choice. The fundamental power to multiply thus becomes a power in a quite different sense, the sense in which human beings are equipped with powers—powers to do so and so, or not to do so and so, as they choose. The conclusions that follow from the theory become correspondingly more open if not always much less harsh. It is no longer a matter of inferring that this or that must happen, ineluctably and unavoidably, whatever anyone may do. Rather, it is a matter of what must happen *unless*, of course, some or all those concerned choose alternative courses.

It is altogether typical of the wayward and uneven track usually followed by the actual history of ideas that Malthus first formulated "the principle of population" as a statement of a physical power, misattributed to our peculiar species; that Malthus later, without really realizing what he was doing, reinterpreted it as the human power which in that context it ought to have been in the first place; and then that, years later still, Darwin, reading the revised version "for amusement,"

revised the revised principle back again to serve as the prime
mover in a great scheme of comprehensive biological theory.
As Darwin put it:

A struggle for existence inevitably follows . . . as more individuals are
produced than can possibly survive, there must in every case be a
struggle for existence, either one individual with another of the same
species, or with the individuals of distinct species, or with the physical
conditions of life. *It is the doctrine of Malthus applied with manifold
force to the whole animal and vegetable kingdom; for in this case
there can be no artificial increase of food and no prudential restaint
from marriage.*[24]

Darwin is thus deliberately leaving choice and the consequences of choices out of account, save insofar as the selective
activities of human breeders of plants and livestock are going
to be that for which natural selection is nature's substitute.
This is no fault at all in his work. But it is the best of reasons
for not expecting to find there any guarantee of constant progress toward humanly agreeable ends. We should also now say
out loud that natural selection is precisely not selection. In
natural selection there are no people or quasipeople choosing,
conscious of the choices being made, and capable at the time
of choosing, of opting in some sense other than that which, in
fact, was or is or will be picked. Indeed, in the original and
still elsewhere ordinary sense of the word "selection," the
expression "natural selection" would appear to be self-contra-
dictory. Of course, there is nothing whatsoever wrong with
employing it in its new and non-self-contradictory interpretation. But the fact that natural selection is not selection is
one more reason for not inferring that anything naturally selected must be, by some independent human standard,
admirable.

[24]Darwin, *The Origin of Species*, pp. 116–7; italics supplied.

Natural Selection as an Invisible Hand

I will not resist the temptation to round off the present section III with a comparison between Darwin's natural selection and Adam Smith's invisible hand. For, just as natural selection is not selection, so this invisible hand is not a hand. Nor, more to the point, is it directed by any intelligence or intention, whether divine or human. The most famous occurrence of the phrase (maybe its sole occurrence in Smith's published writings) is in *The Wealth of Nations*. The context is an argument to show that, as well as why, people investing their own capital within a free market—usually intending only to do the best they can for themselves and their own families—must nevertheless tend to make the most wealth-creating investments possible; thus, as an unintended by-product, they tend to maximize the gross national product. Of the private investor, Smith says:

He generally, indeed, neither intends to promote the public interest, nor knows how much he is promoting it. . . . he intends only his own gain, and he is in this, as in many other cases, led by an invisible hand to promote an end which was not part of his intention.

Seeing this passage, many people—especially perhaps those unaccustomed to hear or even to permit any formidable argument either against state monopoly and command economics or for "the system of natural liberty"—mistakenly conclude that Smith is superstitiously suggesting intelligent and beneficent control by a mysterious and hidden providence. This altogether wrong. Along with his older friend David Hume, with Adam Ferguson, and with several others, Smith was one of the Scottish founding fathers of social science. What these men were doing was to give evolutionary as opposed to creationist accounts of the origin and maintenance of various social institutions, showing these to be unintended consequences of intended actions—actions that were always and only, of course, human actions.

It is perhaps a further paradox in the history of ideas that this work should have been done, without attracting much attention or causing great scandal, in the century before Darwin. For it was in its way very innovative, and it remains very upsetting to argue that some of what are undoubtedly the works of man in fact were not and possibly could not have been products of individual or collective planning and direction. It is not similarly disturbing to argue that plants and animals, which are known not to be human artifacts and are therefore not products of the only intelligence and design with which we are indisputably acquainted, must have developed through the operation of wholly nonconscious and nonrational causes. No doubt everything would have been different had there been some well-regarded argument to design taking off from these social institutions, or had such findings of social science seemed to bear on the question of whether man is a part of nature.[25]

IV. FROM "IS" TO "OUGHT"

The variety of political and moral conclusions that have in fact been drawn from Darwinism, whether rightly or wrongly, is vast. There are also, as we shall be seeing, differences in the ways these conclusions have been drawn. It may be a matter of indicating what is supposed to be an entailment. Or it may be one of suggesting other and logically weaker kinds of connection. Let us start with three crude examples, one of which involves violence, of recommended norms of human conduct being presented as pretty direct deductions from evolutionary biology. Since often nowadays people who employ the expression "social Darwinism" intend it to imply a commitment to what they believe to have been the political economy of Herbert Spencer, it is worth noting, as we review these

---

[25] For more on this aspect of the work of these great Scots, see F. A. Hayek, "Result of Human Action But Not of Human Design," in *Studies in Philosophy, Politics, and Economics* (London: Routledge and Kegan Paul, 1967).

examples, that other and incompatible systems of norms would appear to have an equally good claim to that title.

National Socialist Social Darwinism

To start right at the bottom, we hear Adolf Hitler saying:

If we did not respect the law of nature, imposing our will by the right of the stronger, a day would come when the wild animals would again devour us; then the insects would eat the wild animals, and finally nothing would exist upon earth except the microbes. . . . By means of the struggle, the elites are continually renewed. The law of selection justifies this incessant struggle by allowing the survival of the fittest. Christianity is a rebellion against natural law, a protest against nature. Taken to its logical extreme Christianity would mean the systematic cult of human failure.[26]

Capitalist Social Darwinism

Next, we have the original John D. Rockefeller, who built the Standard Oil Corporation, saying in one of his Sunday school addresses:

The growth of a large business is merely a survival of the fittest. . . . The American Beauty rose can be produced in the splendour and fragrance which bring cheer to its beholder only by sacrificing the early buds which grow up around it. This is not an evil tendency in business. It is merely the working out of a law of nature and a law of God.[27]

Democratic Socialist Social Darwinism

Third, in an editor's preface to E. Ferri's *Socialism and Positive Science*, published in 1905, James Ramsay Mac-

[26] Quoted in H. R. Trevor-Roper, ed., *Hitler's Table Talk* (London: Weidenfeld & Nicolson, 1953), pp. 39, 51. Compare A. Bullock, *Hitler: A Study in Tyranny* (Harmondsworth, England: Penguin Books, 1962), pp. 36, 89, 398–9, 677, and 693.

[27] Quoted in W. J. Ghent, *Our Benevolent Feudalism* (New York: Macmillan, 1902), p. 29.

Donald, who was to become the first Labour Party Prime Minister in Britain, wrote that

> . . . the conservative and aristocratic interests in Europe have armed themselves for defensive and offensive purposes with the law of the struggle for existence, and its corollary, the survival of the fittest. Ferri's aim in this volume has been to show that Darwinism is not only not in intellectual opposition to socialism, but is its scientific foundation.

MacDonald goes on to conclude that "Socialism is naught but Darwinism economized, made definite, become an intellectual policy, applied to the conditions of human society." Ferri proclaims that he is "a convinced follower" not only of Marx and Darwin but also—believe it or not—of Spencer. But he does allow "that Darwin, and especially Spencer [who is elsewhere said to have affirmed aloud his English individualism], stopped short halfway from the final conclusions of religious, political, and social order, which necessarily follow from their indisputable premises."[28] Marx himself never brought Spencer into this reckoning and was, perhaps significantly, less explicit about the supposed relationships. In 1861 he wrote in a letter to Lasalle: "Darwin's book is important, and serves me as a natural scientific basis for the class struggle in history."

Social Darwinism as the Naturalistic Fallacy

The three examples just given all provide textbook specimens of committing the naturalistic fallacy. They all, that is to say, contain or presuppose attempts to deduce normative prescriptions or proscriptions from premises containing only statements of neutral and uncommitted fact. Those who in recent years have taken to denying that the naturalistic fallacy is indeed a fallacy may here be invited to explain either (1) what is the alternative basis of invalidity of these particular

---

[28] E. Ferri, *Socialism and Positive Science* (London: I.L.P., 1906), pp. v, vi–vii, and 1.

arguments or (2) how three valid arguments from the same premises come to yield mutually incompatible practical conclusions.[29]

It has already been pointed out, the first part of section III, that "fitness" is in the context of evolutionary theory defined as "having whatever it may as a matter of fact take to survive." So we cannot validly infer, what Adolf Hitler would clearly have wished us to infer, that the groups which survive in such a struggle for existence must also be rated elites by some other and more exacting criterion than that of mere survival. It is now time to insist, as was not done before, that for evolutionary theory the crux is neither mere survival nor survival at the top of some heap but survival to reproduce— hence the maintenance and sometimes the multiplication of the species. Many sincerely professing social Darwinists have been so misguided as to select for their commendation—as "survivors indicated in the struggle for existence"—putative elites that have, in fact, produced and raised to maturity fewer offspring than the vulgar losers with whom they are thus to be compared. Hitler himself, so far as is known, had no children at all; the largest family among his closest associates, that of Goebbels, destroyed itself in the Götterdammerung of 1945; while the "master race" under their leadership suffered no population explosion on the scale common in this century among the "inferior races" of Asia, Africa, and Latin America.

---

[29] For some discussion both of whether this is a fallacy and of whether it was presented as such by Hume, see Hudson. I will here treat myself to no more than one sharp observation. It is that those who want to answer both questions in the negative are often almost unbelievably inept and naive. Benjamin Gibbs, for instance, in an authoritarian leftist book unpersuasively pretending not to be "a tract *against* freedom," fails to catch Hume's irony. So, he thinks to dispose of him as a spokesperson for the false and foolish thesis that all utterances containing the copula "is" are straightforwardly descriptive whereas all utterances with the copula "ought" are purely prescriptive or proscriptive. Gibbs, *Freedom and Liberation* (London: Sussex University Press, 1976), pp. 8, 116; italics original. Yet the point with both Hume's law and Hume's fork is precisely not that the crucial distinction is always made but that it always should be (Flew, *Philosophy: An Introduction*, pp. 27–28, 112–13).

Another key expression here is, of course, "natural law."
A descriptive law of nature contains as part of its meaning that
whatever is determined by its terms must be contingently necessary, while whatever is incompatible with them must be
contingently impossible. It is, by contrast, essential to the
meaning of a prescriptive or a proscriptive law of nature that
whatever is determined by its terms should be enjoined, or,
as the case may be, forbidden. It is not, I suppose, as essential
to the second sort of law that there should be alternative courses
open to those to whom it is addressed as it is to the first sort
that there must within its scope be no alternative possibilities;
however, the proclamation of a prescriptive or a proscriptive
law loses all point if it is the case that, to a person, its public
will, in fact, obey, necessarily and willy-nilly. It is, therefore,
egregiously perverse to appeal, for instance, to some Hobbist,
supposed psychological law of self-preservation in order to justify a shabby skulking out of danger—perhaps adding, with a
touch of sanctimoniousness, that this conduct is categorically
imperative under the correspondingly Hobbist prescriptive law
of self-preservation.[30] Your appeal to the second sort of law

[30] In chap. XIII of *Leviathan,* a law of nature is defined as "a precept or general
rule, found out by reason, by which a man is forbidden to do that which is
destructive of his life, or taketh away the means of preserving the same;
and to omit that, by which he thinketh it may best be preserved." Compare
this prescriptive and proscriptive formulation with the statement in chap. I
of *De Cive:* "Among so many dangers therefore as the natural lusts
of men do daily threaten each other withal, to have a care of one's self is
not a matter to be so scornfully looked upon, as if so there had been a power
and will left in one to have done otherwise. For every man is desirous of
what is good for him, and shuns what is evil, but chiefly the chiefest of
natural evils, which is death; and this he doth, by a certain impulsion of
nature, no less than that whereby a stone moves downwards."
It is remarkable that so incisive a thinker and so excellent a writer does
not here make it absolutely clear whether what is that we are supposed not
to be able to avoid is having the desires or acting to satisfy them. Insofar
as Hobbes was one of the first of so many to aspire to develop a psychological
mechanics, in which our desires (or "drives") would be construed as forces
compelling us to act toward their satisfaction willy-nilly, we may say that
he slips into any such ambiguity *quasi veritate coactus,* or as if compelled
by the truth(Flew, *A Rational Animal,* chap. VII and passim).

constitutes a tacit admission that the supposed law of the first sort does not in fact hold.

Prescriptions and proscriptions refer to choice. They refer, that is, to those areas of human behavior where alternatives are open to the behaver, who is thus and there truly an agent. But choice is something which, deliberately and with good reason, Darwin leaves out. As was emphasized in section III, it is with this in mind that Darwin stipulates that his fundamental non-Malthusian principle of population must exclude our own species from its scope; as was brought out there, his natural selection is precisely not selection. It should be seen as strange and incongruous that Darwin's theory has so often been seen as providing imperative direction for what he himself was here at such pains not to touch.

Wherever such inferences are made to normative conclusions supposedly entailed by what is in fact a purely descriptive and explanatory theory, the chances are that the term "natural" will play some mediating part. The context of Hume's classical description of the nerve of that invalid form of argument for which G. E. Moore was—nearly two centuries later—to introduce the apt label "naturalistic fallacy" is here significant. This description constitutes the final, ostensibly afterthought paragraph of the first of the two sections of Part I, Book III, of the *Treatise*. Perhaps the most powerful reason for dismissing recent suggestions that Hume did not intend this luminous manifesto to be read in the way here proposed is the fact that, in the second section of Part II, he proceeds to argue, with several illustrations, "that nothing can be more unphilosophical than those systems, which assert, that virtue is the same with what is natural, and vice with what is unnatural."[31]

Hume's reproach goes home against the first two "systems" cited above. For Hitler assails Christianity as "rebellion

---

[31] Hume, *A Treatise of Human Nature* [1739–40], ed. L. A. Selby-Bigge (Oxford: Oxford University Press, 1896), p. 475. And see references in note 29 of the present essay.

against natural law, a protest against nature," while Rockefeller, addressing a Christian Sunday school, takes the same
wrong way to come up with the partially opposite conclusion
that "survival of the fittest . . . is merely the working out of a
law of nature and a law of God." It is also just worth noting
Rockefeller's inept choice of example. The American Beauty
rose, as an achievement of deliberate horticulture, is precisely
not a trophy of Darwinian natural selection; so it belongs here,
if at all, as a specimen of the artificial rather than the natural.

V. SEEING IN AN EVOLUTIONARY PERSPECTIVE

The upshot of section IV is that any attempt to deduce
norms for human conduct from this theory of the origin of
species by natural selection must be as irredeemably wrong-
headed as the search in the same area for some guarantee of
human progress, some assurance of victory for the right. So
far, so negative—and, perhaps, dispiriting. There is, however,
a third way in which evolutionary biology can bear on normative ethics. We can and, I believe, at least occasionally should,
try to see our human activities and inactivities in an evolutionary perspective.

A first example here is that rather reticent sentence from
Marx, quoted once already: "Darwin's book is important, and
serves me as a natural scientific basis for the class struggle in
history." Although there is no reason to believe that Marx was
alert to the error of the naturalistic fallacy, he is apparently
not arguing here, on Hitlerian lines, that since there is a struggle for existence among the plants and the brutes, therefore
people ought to obey the Marxist imperatives of class war. His
point appears to be, rather, that the picture of the history of
the nonhuman living world painted in *The Origin of Species*
is fully congruous with, and hence to some extent parallels,
the theory of affairs at the human level first outlined in the

Communist Manifesto: "The history of all hitherto existing society is the history of class struggles;" and so on.[32]

Fair enough, no doubt; so long as this evolutionary theory is being called on to provide only further support, and much wider application, for the fairly familiar yet nonetheless true observation that it is a rough, tough world. But Marx, as a would-be social scientist, ought to be concerned to discover how findings in other fields relate or fail to relate to his own. This potentially salutary review he cuts off too soon. For suppose that we really do take a good long look at his system in the present evolutionary perspective, regarding it without any Hegelian or other providential prejudices. Then whatever shall we have to think of Marx's utterly confident and never seriously examined conclusions: that this whole history of savage struggles is bound to end—and pretty soon, too—in the annihilating victory of the class to end all classes; and that this final revolution is to be followed in very short order by the smooth establishment of a conflict-free utopia?

Those last words are not too strong. For there can be no doubt that, for instance, the second clause of the concluding sentence of the main body of the *Communist Manifesto* is intended to assert not a tricky tautology but a substantial revelation: "In the place of the old . . . society . . . we shall have an association in which the free development of each is the condition for the free development of all."[33]

Such wild and baseless promises have always been and remain essential to the sales appeal of a product that is, notwithstanding, brazenly marketed as scientific and not, repeat, not, utopian socialism. The denouement in the *Communist Manifesto* is already incongruous enough with the rest of the play. The proletariat in this story will be progressively pauperized: "The modern labourer . . . instead of rising with the

---

[32] K. Marx and F. Engels, *The Communist Manifesto*, trans. S. Moore; ed. A. J. P. Taylor (Harmondsworth, England: Penguin Books, 1968), p. 79.
[33] Ibid., p. 105.

progress of industry, sinks deeper and deeper below the conditions of existence of his own class," and again "The average price of wage labour is the minimum wage, i.e., that quantum of the means of subsistence which is absolutely requisite to keep the labourer in bare existence as a labourer."[34] How could such a battered, harried, exploited, ignorant, wretched mass achieve what has, throughout all the labors of the ages, been beyond the reach of the most fortunate of the wise and good—even given the *deus ex machina* assistance of Dr. Marx and that "small section of the ruling class" that "cuts itself adrift, and joins the revolutionary class, the class that holds the future in its hands"? It is, surely, obvious that on those comparatively rare occasions when Marx speaks of the times beyond he becomes airborne above all history and social science; he becomes, at least for a moment, the revolutionary secular rabbi, the man-intoxicated Hebrew prophet.

So his dreams of perfection are already incongruous enough with the historical materialism which is supposed to constitute their scientific foundation. Yet they are, if anything, even more incongruous with an evolutionary and secular account of human origins. In section II we saw how hard it is to reconcile Darwinian biology with traditional religious assumptions that our species is creation's peculiar favorite. But suppose that you could swallow basic theism and that you were innocent of all knowledge of the work of Darwin and his successors. Then it might perhaps be quite reasonable for you to expect the kingdom of God on earth as the promised consummation of the history of creatures God-designed for that end. You do not pretend to base your confidence in the future on

[34] Ibid., pp. 93, 97. The only gloss I am able to offer on the first of these two statements, with its apparent insistence that the (average) modern laborer tends always to sink below the (average) conditions of the class of which he is a member, is to compare it with a form of utterance which has in recent years become wryly familiar to all students of the British labor unions: "If there is a national average wage increase, then it would be unfair for anyone to have less than the average, though there must of course be some special cases—ours for a start!—in which some group gets a lot more than the average."

that, in your eyes, almost uniformly deplorable history. For you believe that God created men especially to be fit subjects in his peaceable kingdom, and that that deplorable history was all, as it were, an aberration—the result of an unfortunate lapse from the original condition.

Now suppose that, with Marx and Engels, you have, despite your most vehement rejection of all its theological presuppositions, retained a providential scheme—the Eden of primitive communism, the mysterious "fall," the final consummation of history in the restoration of that ancient Eden at a new and higher Hegelian level. At that level, humanity becomes inexplicably exempt from whatever that first fatal weakness was. Then, surely, an opening of eyes to the evolutionary origin of our species should be as fatal to your providential scheme as it is to that of that theological ancestor.

A second example is provided by Julian Huxley. In his younger days, as we saw in section III, he hoped to be able to squeeze out something stronger. But later he seemed to realize that this could not be done. Instead, in his later books, he urged that the correct ways for the moralist and the politician to take account of Darwinian discoveries were quite different. Thus, in the preface to *Evolution in Action,* Huxley maintained: "It makes a great difference whether we think of the history of mankind as wholly apart from the rest of life, or as a continuation of the general evolutionary process, though with special characteristics of its own," for "In the light of evolutionary biology man can now see himself as the sole agent for further evolutionary advance on this planet. . . . He finds himself in the unexpected position of business manager for the cosmic process of evolution."[35]

Two questions now arise: Does it indeed make a great difference? And does such a way of seeing the human situation have any claims on anyone who has not enjoyed a biological training? What is to be said against the impatience of Jane

[35] J. Huxley, *Evolution in Action* (London: Chatto and Windus, 1953), pp. vii, 132.

Welsh Carlyle? Mrs. Carlyle, it may be recalled, confessed that she "did not feel that the slightest light could be thrown on my practical life for me, by my having it ever so logically made out that my first ancestor, millions of millions of ages back, had been, or even had not been, an oyster."[36]

The first point to make in reply is that the evolutionary perspective takes for granted various general propositions that are in fact true. There truth often bears on that of others which do play crucial parts in various conduct-guiding world outlooks. This, as we have already seen, is one excellent reason why both the atheist Marx and the protagonists of traditional Christianity needed to ponder the implications of seeing our lives in an evolutionary perspective.

Or consider a third case, involving in the first instance metaethics rather than ethics. In that curious, extraordinarily parochial volume *Principia Ethica,* the whole discussion proceeds as if outside time and space. Moore and his culture circle appear never to have heard travelers and historians tell of people cherishing very different values. The argument is equally undisturbed by any news from modern science. Moore might as well have been writing not merely before Darwin but before Newton.

Take, for instance, Moore's treatment of the naturalistic fallacy and compare it with that of Hume. (Hume, incidentally, is not mentioned in Moore's book.) Had Moore seen man as a part of nature and in an evolutionary perspective, how could he have gone on as he did about the detection of his strange, inexplicably nonnatural characteristics? How could he then have failed to conclude, as Hume had concluded nearly two centuries earlier, that value judgments, which are not deducible from any neutral propositions referring only to what is the case, must instead involve some kind of projection of individual or collective human activity or desire, much as Newton and Galileo had concluded that the so-called secondary

---

[36] J. W. Carlyle, *Letters and Memorials,* vol. III (London: Longmans Green, 1883), pp. 20–21.

qualities are projections of what is really only "in the mind" onto things in themselves, the only authentic qualities of which are primary?

The second and third points in response to the formidable Mrs. Carlyle can be dealt with more briefly, though this is no indication of lesser importance. Both are familiar justifications for "taking a wider view," though again not on that account to be dismissed as inconsiderable. One is that such wider views may enable us to see things that do not emerge so easily, if at all, from a more blinkered appreciation. This point can be well illustrated by the case of Julian Huxley himself. For it was precisely an evolutionary vision that determined his own recognition, long locust years before this was even as widely admitted as it is now, that drastic checks on human fertility are a necessary condition for the maintenance of humanity's estate—to say nothing of its relief!

The other remaining point is that many people long to see things as a whole, to find some deep, comprehensive, world picture against which they may set their lives. No philosopher has any business either to despise or not to share such yearnings. Here the Darwinian vision possesses the neither universal nor despicable merit of being founded upon and not incompatible with known facts. Also, as Darwin said in that final peroratory paragraph of *The Origin:* "There is a grandeur in this view of life. . . ."[37]

I can think of no better way of ending this paper than by recalling the view expressed by Julian Huxley in *Evolution in Action:*

In the light of evolutionary biology man can now see himself as the sole agent of further evolutionary advance on this planet. . . . He finds himself in the unexpected position of business manager for the cosmic process of evolution. He no longer ought to feel separated from the rest of nature, for he is part of it—that part which has become conscious, capable of love and understanding and aspiration. He need no longer regard himself as insignificant in relation to the cosmos.[38]

[37] Darwin, *The Origin of Species*, pp. 459–60.
[38] Huxley, *Evolution in Action,* p. 132; cf. Flew, *Evolutionary Ethics*, p. 60.

CHAPTER TWO

Evolution and Ethics

ERNST MAYR

Professor Flew has made my task very easy, because his presentation of the Darwinian theory reflects quite closely the best modern thinking. There are a few points on which I disagree with him, and there are others that he chose not to deal with but which I hope to go into.

Let me get the disagreements out of the way first. I want to take strong exception to Wittgenstein's pronouncement that "The Darwinian theory has no more to do with philosophy than has any other hypothesis of natural science." This is a typical example of the ignorance of the Vienna circle and their total inability to understand anything that has to do with biology. Wittgenstein and all others that ultimately can be traced back to the Vienna school (including, until quite recently, even Karl Popper) were strongly opposed to Darwinism because Darwinism completely refuted essentialism, which, after all, was the major basis of the thinking of the Vienna school.

Actually Darwin created a major revolution in philosophy, even though he never wrote a book entitled "The New Evolutionary Philosphy." In refutation of Wittgenstein, let me ask: What other theory in the natural sciences is as rich in brand new philosophical concepts? I mean this in the strict technical sense of philosophy. It is only in the last twenty-five years or so that the younger philosophers have begun to incorporate these basic new concepts in their thinking. I refer to such concepts as the following:

1. The population concept, with its emphasis on the unique-ness of each individual and the total absence of any basic class norm or essence. I might mention that the mathe-maticians are still quite unable to cope with populational thinking.
2. The duality of the selective process, consisting of a first step producing inexhaustible variability (a step almost en-tirely controlled by chance or accident) and a second, di-rection-giving step consisting of a probabilistic-statistical ordering (a strong antichance factor). To the best of my knowledge, no similar concept existed previously in philosophy.
3. The demonstration, as Scriven has rightly pointed out, that prediction is not an inseparable part of causality.
4. The idea that teleology is an unfortunate cover name for three or four entirely independent phenomena, some of which have a strict chemicophysical basis while others do not.

The writings of Beckner, Scriven, Hull, Munson, Wimsatt, and others have demonstrated what a revolution in strictly technical philosophy the application of these Darwinian con-cepts is producing—a revolution that, at the present time, is only in its early stages.

It is owing to their adherence to the totally inappropriate concept of the physical sciences that so many philosophers are having such a difficult time in applying Darwinism. It is unfortunately necessary for the philosopher to keep up with the steady advances in the thinking of evolutionists in order to apply the results of these advances to general philosophical theory.

One result of the modern analysis of Darwinism is that the so-called Darwinian theory is actually a highly composite *paradigm*, to use this word in Thomas Kuhn's sense, consist-ing of at least five separate theories. Some of these theories do and some others do not have ethical implications. Let me skip the two Darwinian theories that have no obvious ethical implications (they relate to the origin of diversity and to the gradualness of evolution) and turn to the three Darwinian theories of ethical relevance.

The first of these theories proposes to replace the constant world of Christianity with an evolving world. According to Christian metaphysics, the earth was created only a short time ago. Bishop Usher calculated that creation had taken place about six thousand years ago and furthermore that the world would find its end on a day of judgment that was not far away. Man, in such a constant world of short duration, had no obligation to the future, not only because God had told him that he had created the world and all of its creatures strictly for his benefit but also because God was controlling everything in this world anyhow.

Man had obligations only to God and to his fellow man. This is why Christian ethics is so extraordinarily individualistic and man-centered. The ethical norms were directly revealed to man by God—for instance, in the Ten Commandments— and God's word not only provided a firm basis for the solution of all ethical problems but did so in an unequivocal manner.

Having the world evolve, and to make matters worse by materialistic means—that is, by natural selection—deprived man's morality of its very foundations, as was bitterly decried by Adam Sedgwick and other natural theologians who opposed Darwin. They challenged the Darwinians to demonstrate the possibility of an evolutionary basis and standard for morality, and the two Huxleys and numerous other evolutionists have been busy ever since in efforts to establish such a foundation.

In many ways it will have to be a considerably different set of ethical norms from the Christian one. Evolutionary biology and other sciences, like geology and cosmology, have shown us that the constant world of short duration of Moses and the Bible does not exist. Not only is the world not constant or of short duration, but man himself has an immense power to change the world, of which he himself is an evolved part. It is perhaps the most important impact of the theory of evolution that is has expanded man's vision from the parochial world view of a Near Eastern pastoral society to that of a man who is conscious of his evolutionary past and aware of his responsibility to posterity. Modern man's challenge is to develop an ethic that can cope with this new vision.

The admonition "Love they neighbor" is no longer sufficient; the new ethic must include a concern for the community as a whole and for posterity. What makes the development of such a revised set of ethical norms so difficult is that the duties toward oneself, one's neighbor, the community, and posterity are often in conflict with one another or even completely opposed. This is why we have such a phenomenon as the tragedy of the commons.

Let me now take up the ethical consequences of another Darwinian theory, the theory of common descent. This theory deprives man of his unique position in the world, refuting the biblical view that the whole world was created for the benefit of man and that man has a complete license to do anything he likes with nature. Such a man-centered attitude flies, of course, straight in the face of any conservation ethic, as Lynn Whyte has told us.

By contrast, much of the modern conservation ethic is an inevitable by-product of the theory of common descent. Every creature is a unique and irreplaceable product of evolution, and man has no right whatsoever to exterminate even the least of them.

The theory of common descent raises another and indeed most intriguing problem: namely, that of the origin of the human ethic. How much of man's ethic is a heritage of his primate ancestry? Many philosophers would at once object to this question, contending that ethics requires consciousness and other traits that, they claim, are found only in the human species. But is this really true? I know quite a few biologists who are convinced that even a paramecium or an amoeba has a primitive consciousness. Consequently we must set aside all preconceived notions and ask the completely uncommitted question as to whether there is any evidence for the existence of ethical norms in animals. Lack of time prevents me from following up this question, but every dog owner knows that dogs, when committing a forbidden deed, display all sorts of guilt feelings, demonstrating thereby that they have a clear perception of good and bad. They also are capable of acts we would normally designate as altruistic, as I shall discuss presently.

Similar observations have been made for anthropoid apes, and it can hardly be doubted that something is present in many animals that can be considered the raw material of man's ethical norms. But how was the giant step from this rudimentary basis to man's ethical system achieved?

I entirely agree with Professor Flew that evolution does not lead automatically to a "higher" living being. Even though there is a rough progression from the lowest and simplest organisms to the highest plants and animals, this progression is not the product of any built-in drive toward perfection, as believed by Lamarck, or of the existence of any teleological principles. Whatever progress occurred during past evolution was not a programmed progress but rather the uncertain and haphazard progress inherent in any production of natural selection. I agree with Professor Flew that it is impossible automatically to deduce norms for human conduct from the theory of natural selection.

As necessary as it is that we have a correct understanding of the Darwinian basis of evolution, it fails to tell us how high ethical principles and altruistic behavior could have evolved. To get a little closer to a possible solution we must now try to analyze Darwin's fifth theory of evolution, the theory of natural selection. If we look at natural selection in terms of the thinking of Herbert Spencer and the sociologists as a fierce struggle for existence, we will not get anywhere. To be sure, the concept and the terminology of struggle for existence existed long before Darwin, but—as is by now forgotten—it was for the natural theologians a struggle among species, of the wolf against the deer and the cat against the mouse.

The great insight that Darwin had on September 28, 1838, was that the only struggle of relevance for evolution is that among individuals of the same species. How could such a struggle affect behavior and the norms on which behavior is based? At first sight one would think that only manifestly selfish behavior would benefit an individual (i.e., would be of selective advantage). And indeed, there is much evidence for an ingrained tendency toward selfish behavior, and a great deal of education consists in suppressing this. However, to think of evolution only as a struggle for existence, in the sense of the social Spencerians, misses the mark to such an extent that it is of no explanatory value at all. A totally individualistic ethical philosophy simply does not work with a social organism such as is man.

The trouble lies with the metaphor "*struggle for existence.*" Darwin himself is partly responsible for this, since he entitled the third chapter of the *Origin* "Struggle for Existence." While using this traditional term, however, he made it very clear that his own concept was far more sophisticated. As he states in the *Origin*, "I should premise that I use the term Struggle for Existence in a large and metaphorical sense, including dependence of one being on another, and including (which is more important) not only the life of the individual, but success in leaving progeny." In this sentence Darwin laid the foundation of two major developments in evolutionary biology, one based on the words "dependence of one being on another" and the other on "success in leaving progeny."

The words "success in leaving progeny" make it quite clear that natural selection is not simply "survival of the fittest" but rather differential reproduction. Such success in reproduction, Darwin thought, could be due to two entirely different sets of properties. One of them Darwin called "natural selection" in the strict sense of the words, and the other "sexual selection." The two cannot always be neatly separated, but there are clearly characteristics that improve a species as a whole, such as improved temperature tolerance, an improvement in the sensory organs (vision, hearing, etc.), better predator escape mechanisms, and so on. By sexual selection Darwin understood a selection for any personal characteristics that enhanced the chances of reproductive success of the individual without benefiting the species as a whole, as, for instance, when male birds of paradise acquire more gorgeous plumes and thus attract more females in these non-pair-forming species. Darwin's theory of sexual selection was wrong in many minor details, but it was quite right in its clear recognition that the means by which reproductive success is achieved very often do not benefit the species as a whole. Curiously, as Hamilton and Trivers have shown, it is the selfish propensity to increase the reproductive success of one's genotype that is responsible for the unselfish behavior of the workers in the colonies of social ants and bees.

With the statement, "dependence of one being on another," Darwin had already pointed out that altruistic behavior might be of selective advantage. Haldane took this idea up in 1932, and calculated how closely related two people would have to be to make it advantageous, for the reproductive success of a particular genotype, for one of them to risk his or her life on behalf of the other. Hamilton and others have worked this out in far greater detail, and this branch of the theory of natural selection is now called the "inclusive fitness" theory. Williams, Trivers, Alexander, and others have made the major contributions to it. There is still much controversy as to details, but the major finding, which has not been refuted, is that altruistic behavior is quite often of high selective advantage when it enhances the reproductive success of a given genotype. Obviously, such behavior is of selective advantage only if it is directed toward close relatives. However, in a social organism such behavior may lead to a strengthening of the cohesion of the social group and may thus be of selective advantage for all members of the social group, even beyond the point where Hamilton has calculated that relationship would bring reproductive advantage. It is at this point that Triver's theory of reciprocal altruism sets in, which states that there is a selection for a type of behavior that might facilitate a "give and take" of mutual favors, even beyond the narrow confines of immediate relationship. I shall not go into details because this theory is still controversial; furthermore, it leads us into a dangerous borderline zone between individual and group selection.

The concept of inclusive fitness is one of the recently developed evolutionary concepts that is indispensable for explaining the relations between ethics and evolution. Another is the concept of personal investment. It is likewise based on the idea that a genotype wants to perpetuate itself as much as possible through reproductive success. It was Bateman (1949) who pointed out that there is a pronounced difference between males and females in reproductive investment. There is a great expenditure for females in developing eggs, in often retaining them until the embryos reach maturity for independent life, and so on. The number of offspring a female can produce is quite limited. Since there is no pair bond in the majority of species of animals, males can fertilize numerous females, and the metabolic cost of producing spermatozoa is quite minimal. This leads to totally different behavioral strategies for males and females and is the cause of many behavioral norms. It also explains parent–offspring conflict and many other behavioral phenomena that have ethical significance. Robert Trivers, in particular, has recently studied the numerous ramifications of the principle of personal investment and parent–offspring conflict. The investment concept is also involved in the principle of reciprocal altruism. This latter concept is particularly important, because almost all the higher ethics—such as friendship, gratitude, trust, sympathy, and so on—can be derived from it.

Ethics means ethical behavior. The explanation of behavior, its genetics, and its physiological and evolutionary causation is the key problem in the relation between ethics and evolution. It is unavoidable that at this point we have to take up once more the twin problems of how much of a genetic component ethical behavior has and how evolution can incorporate this genetic component into the genotype. What is important (and unfortunately has often been forgotten in recent debates) is that the ethical behavior patterns that are involved in the evolutionary phenomena I have just described—such as inclusive fitness, parental investment, and reciprocal altruism—are in part controlled by open behavior programs. Since the concept of closed and open genetic programs is relatively new, I presume I must add a few explanatory words.

Let us take the simplest case of a strictly instinctive behavior pattern of some lower animal. When it receives the appropriate stimulus, it performs the appropriate behavior. The behavior in this case is entirely controlled by the genetic program, and we call a genetic program that leads to such stereotyped, unmodifiable behavior, a "closed program."

Almost always in higher animals, some if not most behaviors are also influenced by previous experiences, that is, by information stored in the memory. We say that such behaviors are controlled by an "open program."

We have learned a great deal in recent decades as to which behaviors in what organisms are controlled by open or closed genetic programs. Species with a short life span (e.g., as in the extreme case of mayflies where the adults live only a single day) have no opportunity at all to store up experience and have therefore been selected to have all the information they need in their short adult life stored in their genetic programs. The longer an organism lives, particularly if it has parental care during which information can be transferred from the parents to the offspring, the more selection favors an open program in which new information can be stored. Some of the behavior even in these species is closed, or is strictly genetically controlled. The term *open program* means that there are, so to speak, compartments in the central nervous system that are ready to store new information.

When we consider the evolution of the primates, that is our ancestors, we find that there is a more or less well-defined phylogenetic trend, from the lowest prosimians to the anthropoid apes, for an increase in brain size, which, in turn, is closely correlated with a replacement of closed by open behavior programs. However, there is no other organism in which the behavior program is as open as it is in man. Most of our behavior is guided by norms that were acquired after birth by education or self-learning.

It must not be forgotten, however, that even in man there are behavioral tendencies that involve a genetic component. Furthermore, the open program itself has a genetic framework that facilitates the integration of acquired information with inherited information. The magnitude of this genetic component is documented by the inherited brain size.

We know exceedingly little about the properties of open programs, but we do know two things that are of great importance for the ethicist: (1) Certain pieces of information are incorporated more easily than others, and (2) There are periods in life, particularly in youth, when certain types of information are incorporated more easily than at other times. It is this second insight to which Waddington has particularly called attention, and I think with full justification. In his book *The Ethical Animal* (meaning man) Waddington postulates that part of the brain is set up to accept and store ethical norms by a process the ethologists had earlier described as "imprinting." This explains why certain ethical principles acquired in childhood are often so firmly adhered to for the rest of the individual's life. It is this same open program that is responsible for the relatively high success of much of early religious indoctrination, leading to the maintenance of beliefs which seem to other persons who are not so indoctrinated to be refuted by common sense or by science.

It is the possession of such an open behavior program and the fact that it is being filled in childhood or youth which accounts for the human capacity to have ethical norms. It also tells us why whatever ethics animals may have are really quite different from ours. What this theory of Waddington's does not explain is the development of the particular ethical norms that are incorporated into the open program.

It is at this point that we must remember that man, although evolutionarily speaking an animal, is a very special, indeed unique, animal. He is distinguished from any other animal by numerous characteristics, among which, in this context, the most important is the ability to develop culture and to transmit it from generation to generation. Man's ethical norms are part of this cultural tradition, and this explains why there are often such astonishing differences in ethical norms between different human cultures and ethnic groups.

For those who are strict believers in some particular creed or religion, it is usually easy to know what is right or wrong because it is codified in some sort of sacred writings or revealed messages. For those others who are agnostics, the task is more difficult. How are they to know what ethical norms to choose? Here I go a long way with Julian Huxley, who said that every thinking person has a religion, even if it should be a religion without God. And for the evolutionist, that religion is what Huxley calls "evolutionary humanism." It is a belief in man, a feeling of solidarity with mankind, a loyalty toward mankind. Man is the result of millions of years of evolution, and our most basic ethical principle should be to do everything toward the maintenance and future of mankind. All other ethical norms can be derived from this baseline. If we believe in this, we can see at once that in addition to the individual ethics preached by Christianity, there has to be a community or social ethics.

Again and again in the last one hundred and fifty years the mistake has been made of treating man as "nothing but an animal." I emphatically agree with Professor Flew in his refutation of this viewpoint, which was promoted by certain radical writers as well as by some of the popsociobiologists. It is particularly important for those who are not biologists to realize that these popular writers do not at all represent the views of the vast majority of evolutionary biologists.

Evolutionary humanism, if properly understood, can serve as the basis of an eminently practical as well as eminently satisfying ethical system. It is a demanding ethics, because it tells every individual that he has somehow a responsibility toward mankind, and that this responsibility is or should be just as much part of his ethics as individual ethics. Any generation of mankind is the current caretaker not only of the human gene pool but indeed of all of nature on our fragile globe.

If we want to summarize our findings, it is that man's ethical outlook is, or at least should be, fundamentally affected by evolutionary theory. Evolution does not give us a complete, codified set of ethical norms such as the Ten Commandments, yet an understanding of evolution gives us a world view that can serve as a sound basis for the development of an ethical system that is both appropriate for the maintenance of a healthy human society and that also provides for the future of mankind in a world preserved by the guardianship of man.

CHAPTER THREE

Darwinism and Ethics

A Response to Antony Flew

LEON R. KASS

What bearing should Darwinism (that is, the theory of the origin of species, including the human species, by natural selection) have on ethics (that is, on our thoughts about how we as human beings should live our lives)? This question is but a recent and specific example of the age-old and general question about the relation between our knowledge of nature and our thoughts about ethics. This broader question, in turn, presupposes prior questions, both about the nature of nature and about the nature of ethics and the good for man. A full exploration of the former would carry us into questions about change, time, energy, cause, and, ultimately, about *being* itself. A full exploration of the latter would involve questions about justice, nobility, freedom, virtue, duty, happiness, pleasure, and, ultimately, about *good* itself. Both explorations would also necessarily consider the intelligibility of what *is* and what is *good*, as well as the powers and limits of the thinking mind to *discover* the intelligibly true and good. In short, the question with which we began takes us quickly into an inquiry about the whole, an inquiry whose passionate pursuit has always been the true activity of philosophizing, of the seeking after wisdom about the whole. Neither Professor Flew nor anyone else should be faulted for failing to address all these questions in a brief paper, though I confess I wonder whether Professor

Flew's professed self-understanding of philosophy[1] and his own display in this paper of how he "does philosophy" permit him to reach and think fundamentally about these matters.

Though I shall not be able to take up these most crucial questions, I do wish, by means of a few preliminary and general observations about the relation between our knowledge of nature and our thoughts about ethics, to open the way for their consideration.

I. NATURE AND ETHICS

To inquire properly into questions about the relation between nature and ethics requires, to begin with, that we liberate these questions from the tyranny of the logicians. That relation may be much richer than is suggested by its treatment as a problem of the logical connection between "is" statements and "ought" statements. Does the fact that logic distinguishes the mode of description and the mode of prescription necessarily imply that there is no connection between *what is* and *what is good*? Is the fact–value distinction, upon which the is–ought distinction rests, sound—and, by the way, does it state a fact or a value? Professor Flew apparently thinks it *is* sound, as he apparently endorses the conclusion that

value judgments, which are not *deducible* from any *neutral propositions* referring only to what is the case, *must instead* involve some kind of *projection of individual or collective human activity or desire*, much as Newton and Galileo had concluded that the so-called sec-

---

[1] For example, in his opening paragraph, Professor Flew asserts that most of Plato's *Laws* and much of the *Republic* "is not philosophy" except in the loose sense that opinions about general matters constitute, say, my "philosophy" of life. Would Plato, or any other genuine philosopher in the tradition, agree? Later, he makes the peculiar remark, "No philosopher has any business either *to despise* or *not to share* such longings," that is, longings "to see things as a whole, to find some deep, comprehensive, world picture against which they may set their lives" (p. 33). One would have thought such desires were the heart of the philosophical activity itself. Does any other kind of activity deserve the name of philosophizing?

ondary qualities are projections of what is really only "in the mind" onto things in themselves, the only authentic qualities of which are primary. (pp. 32–33, my emphasis)

Is it true that our judgments *must be* projections of our "activity or desire" (rather than discoveries) and *must be* projections *because* they are not deducible from "neutral propositions referring only to what is the case"? Is "what is the case" simply "neutral"? And is even our knowledge of the nature of things strictly *deducible,* and from "neutral propositions"? What is Professor Flew's or our answer, on the one hand, to Kant, who argues that the understanding *prescribes* to nature its laws (with the further implication that nothing is "known" in itself with greater certitude than the *moral* law); and to Nietzsche, who argues that all "truth" is altogether a human *creation;* or, on the other hand, to Aristotle, who seems to suggest, correctly in my estimation, that some people can indeed *directly* and accurately discern—by the mind's eye—a noble or just action, in the same way that most of us correctly discern "green" or "two"? Is it not possible that some excellences of character and fineness of deed simply show themselves forth to the minding eye, carrying the evidence of their goodness in themselves? Is what we call "good" good only because we value it, or do we value it because it is good? (Even Professor Flew appeals [p. 15] to our intuitive awareness of "*splendid*" physical or mental endowments, "*wretched* creatures," "*superb* animals," "*genius,*" and, several times, to the "*grandeur*" that is apparently *evident*—and not merely *projected*—in Darwin's view of life.)

The naturalistic fallacy is indeed a fallacy (i.e., an error in *logic*). A fallacy is an error, but an error only in *reasoning*. Some conclusions, wrong *as conclusions* because of faulty logic, might nevertheless tell the truth *about the world*. Thus, the implied assertions that fact and value, or better, the true and the good, are separate and nonoverlapping realms, and that "good" is a projection of desire rather than an insight into "what is the case," require separate arguments—arguments not merely about "statements."[2]

But even if the true and the good do not belong to wholly separate universes (e.g., to so-called objective and subjective realms), there are good reasons for not being overly sanguine about discovering their relation and better reasons for suspecting that relation to be of limited value for ethics. This is especially true if one holds that ethics is primarily a matter of rules—"prescriptions and proscriptions refer[ring] to choice" (p. 27)—for no one holds that such rules can be simply "read off" from the natural record. Aristotle, who thinks about the good of man in relation to what he takes to be his natural work, within the natural whole, and who indeed distinguishes something he calls the "just by nature" from the "just by convention," "deduces" from man's nature no *rules* of conduct.[3] In fact, in all of Aristotle's *Nicomachean Ethics* there are only one, or at most two, prescriptions (more accurately, exhortations) and not a single proscription. For Aristotle and for other thinkers who look to nature, it seems that ethics is not so much a matter of rules of conduct as of ways of life (the Greek word *ēthos* means character, or disposition, or accustomed way), virtue or excellence being the perfection of human possibility, cultivated through proper habituation and thought, toward which

---

[2] Professor Flew, against his own strictures, often appears to commit the equivalent of the naturalistic fallacy, drawing conclusions about what we *should* believe or hold from his descriptions of what some men *do* believe or hold: (1) "The two understandings of philosophy thus illustrated *should* both be acceptable" (p. 4; ironically, this remark is but one sentence removed from his insistence on the paramount importance of the "*ought–is*" question); (2) "There *is* indeed 'a grandeur in this view of life,' and it is something that *should* be part of the world outlook of every modern person" (p. 17; Why *should* it? Because it is grand? Modern? True? Why, according to Professor Flew, *ought* we to believe what is?); (3) "He [i.e., man] no longer *ought to feel* separated from the rest of nature, *for* he is part of it" (p. 33; Professor Flew is approvingly quoting Julian Huxley). Apparently Professor Flew believes that certain kinds of sentiments or judgments are not just arbitrary projections but *fitting responses* called forth by the way things are.

[3] Indeed, Aristotle explicitly denies that there are *any* universally valid rules of action, and says, quite enigmatically, that all of justice—including natural justice—is changeable. (*Nicomachean Ethics* 1104a4, 1134b30)

we are innately but *weakly* pointed and disposed but from which we will readily go astray in the absence of proper rearing.

In tacitly questioning whether ethics is primarily about prescriptions and proscriptions, about "oughts" and "ought nots," I also mean to ask whether it is true, as Professor Flew intimates, that mortality is primarily a matter of choice among alternatives, rather than, say, a matter of *eros*. Choice (*proairesis*) for Aristotle is not primarily a weighing of alternatives but of finding means to ends, and he denies that we *choose* the ends. Is our so-called "choice" of a way of life not better represented by the image of the lover beckoned by or drawn to the beloved than by the image of the judge weighing on a balance? Is not the heart of education learning to love and hate the right things?

Even for an ethics of virtue, the relation of ethics to nature is hardly simple. There is the notorious difficulty in discovering the perfection of the human. The mention of the matter of rearing reminds us of law and custom, culture and politics. Though all men may share the same nature, man is by nature the animal that lives by culture or convention. Cultures vary, and likewise ideas of the noble and the just. Though far from proving that the noble and just are strictly relative—indeed, *that* every culture thinks *something* about justice is *the* warrant and invitation to seek after what is *truly* just—the multiplicity of cultural ways makes difficult the identification of both the naturally human and the humanly best, especially as each investigator faces the difficult though not insuperable task of freeing himself from the conventions of thought imposed by his own rearing (e.g., for us, the fact–value distinction, or our absolute insistence that "there are no absolutes"). And even were we able to discover the natural perfections of the human, we would still be left the difficult work of both unraveling in thought and weaving together in practice the proper complex pattern of nature and culture. For conventions are indispensable for the cultivation and expression of the natural and the naturally best, just as the learning of a particular language (itself strictly a convention) is necessary for realizing our natural human capacity to speak and think. The unavoidable place of convention and opinion in our formed views of the noble and the just raises the further difficulty of whether ethics can or *should* escape from the constraints of particularity and the parochial; that is, whether the contemporary prejudice of ethicists, following Kant, in favor of universality does not unreasonably make too much of *rationality* (noncontradiction) by abstracting from our necessary (and desirable?) particularizing attachments to body, place, and custom.

But above and beyond these difficulties, it is not obvious that knowledge of the nature of nature should or must play a decisive role in thinking about human virtue. What have physics and metaphysics to do with ethics? Do the virtues of Aristotle's *Ethics* cease to be virtues—does cowardice supplant courage or profligacy moderation—just because Aristotle was wrong in attributing life to the heavenly bodies? Indeed, men with opposing views of nature sometimes share the same view of human good. For example, both Aristotle and Lucretius argued that the philosophical and contemplative life was best, though the former thought the cosmos finite, intelligible, lovable, deathless, and a place in which man could truly be at home, while the latter held, much as does modern science, that the whole is infinite and ultimately unknowable, that everything humanly lovable is perishable while everything imperishable (atoms and the void) is unlovable, and that man himself and his world are only temporarily on the cosmic stage, in a drama directed by forces absolutely indifferent to human welfare.[4] Conversely, the same cosmological belief may go along with radically different views of how to live. A roughly Lucretian view of the nature of things has led modern Western man not to contemplation but to action and art, either to the scientific project for the mastery and possession of nature, to make her more responsive to human desires, or to bold, "authentic," creative acts of self-assertion against the prospect of cringing despair and angst about nothingness. Regarding even

---

[4] It is true that Lucretius gave politics and civil virtue much lower marks than did Aristotle.

the most general matter of a way of life, it seems, there is no simple "deduction" from the nature of nature. This means, to anticipate, that the way of life shown forth in the Bible may not be so easily felled by the theory of natural selection, even were it to be the whole truth about the origin of species.

And yet, notwithstanding all these caveats, the nature of nature cannot be irrelevant to our thoughts about ethics. In thinking about human life, it surely matters—or ought to matter—to us what sort of a place the universe is, what its source and highest principles are, and how hospitable it is to human aspiration; in short, how it stands between man and the whole, in which he is at worst a neglected alien and at best a beloved citizen. Closer to home, and also central both to questions about the nature of nature and the human good, is the question about *nature of man*. All accounts of the good life for man explicitly or tacitly presuppose or rest on, at least in part, some understanding of what man is. We cannot think intelligently about how we are to live unless we know our nature—what we are, of what we are capable, and what is good for us. Accordingly, I turn to the more immediate subject of Darwinism and ethics by asking about the bearing of Darwinism on our understanding of the nature of man.

II. THE NATURE OF MAN

Professor Flew suggests that Darwinism requires important revisions in our "estimates of the nature of man," that it makes "plumb unbelievable" the assumption that man has a "special status" in the whole, that it "must tend to discomfit adherents of [should we read "must make unbelievable"?] Platonic–Cartesian [a mixed marriage, to say the least!] pictures of the nature of man." Is he right? Assuming that it is simply true that man arose from nonhuman ancestors by "entirely natural" processes, what difference does this make for our understanding of *what man is*? It is worthwhile distinguishing several matters: (1) the importance of knowing our origins for a knowledge of man; (2) man's relation to other animals in terms of uniqueness, rank, and rule; (3) the end of man; (4) man's relation to evolution; and (5) man's relation to the divine.

General Remarks about Origins

To understand *what* something is (including what it might be for) would seem not to require, necessarily, knowing how it got to be that way. For instance, "What is man?" and "What is man's work or function?" are questions that can be considered independently of man's origins. (This is a slight exaggeration, since if man has been *designed*, his *telos* might have been given him by his designer as well.) We learn what man is first, and also finally, only by studying him as he exists, through and beneath his many cultural guises. The "what" of man—if indeed he has a "what" (see below)—is in the most important respects independent of the from-what out-of-which he came or comes.

We come to know man, and each man, from his looks, his deeds, and his speeches; through these, we also learn of his desires, his passions, and his thoughts. We know man, and each man, also from his associations, from the domestic to the political, yes, even from the religious. To such knowledge of what man is, gained from the human study of the functionally human, the sciences of anatomy and physiology, biochemistry and genetics provide supplements, supplying the knowledge of materials, parts, and mechanisms involved in man's being what he is. But although what man is *requires* these materials and mechanisms, the "what" of man is not identical with them. (This holds even if the "what" or nature of man is itself subject to change: the argument here, an argument about parts and wholes, does *not* depend on man's immutability.)

The same holds true about the temporal "origins" or (better) antecedents. The life of a butterfly cannot be deduced from or explained by the life of its caterpillar or that of a sea urchin from its free-swimming pluteus larva. Moreover, it is possible for something to originate for one reason, or even for no reason—that is, by chance—and yet continue to be and to function for another reason.[5] Powers like intelligence, selected for their contributions to survival, might acquire a life of their own and pursue activities and goals that have little to do with survival (e.g., mathematics or painting or discovering the origin of species). In terms of natural selection theory, to act *as if* for the sake of something and to act *in fact* for the sake of something might have identical selective advantage, even though that something did not govern the original coming to be.

If this is true, then it may make little difference whether man came to be gradually out of the slime or whether he has always been, whether he was formed by an intelligent maker or whether he came together largely by chance. Man would still *be* what he *is,* even if the first men arose through a coalescence of arms, legs, torsos, and heads that were raining separately from the skies. The study of man's origins tells us very little about his nature that cannot be discovered from the study of man himself—considered, of course, in relation to the rest of what now is.

Perhaps this is an exaggeration. Certainly, to understand the character of individual men, it is often important and useful to know their roots. What someone is may very much be influenced by who his parents were. Still, here too the character of a man is not simply deducible from the source. Homer identifies his heroes both by their patronyms and by their epithets—Odysseus is son of Laertes and also Sacker of Cities or resourceful Odysseus, but one cannot deduce "Sacker of Cities" or "resourceful" from "son of Laertes." Still less can this be done where an ancestor is merely the antecedent progenitor and not the *source* of one's rearing in character, or even of one's untrained nature, as is the case under the theory of evolution when we look at man in relation to his nonhuman

---

[5] For example, "the polis comes to be for the sake of life, but it exists for the sake of living well." (Aristotle *Politics* 1152b29). Another example: "[S]uch 'means' of survival as perception and emotion are never to be judged as means merely, but also as qualities of the life to be preserved and therefore as aspects of the end. It is one of the paradoxes of life that it employs means which modify the end and themselves become part of it." (Hans Jonas, *The Phenomenon of Life* [New York: Dell, 1966], p. 106.)

forebears. Indeed, it is not unimportant to note the difference between an origin or source which is the *cause* of the *nature* of the being that it engenders and a so-called origin that is merely its precursor or antecedent. The latter has little causal *responsibility* for the *what* of its descendants. (In this connection, we note in passing that Darwin's book should have been entitled *The Genesis*—or coming to be—*of Species;* it teaches us nothing about their sources, any more than it teaches us their nature.)

Yet this too is perhaps an exaggeration. Two difficulties at least are worth mentioning, though I cannot consider them both here. First is the challenge of the philosophy of history— associated with a tradition from Rousseau, through Kant, Hegel, and Marx, to Heidegger—which advances on the strictly human plane certain theses like those put forth by Darwin about living nature as a whole. Is it really true, as is now widely claimed, that man has no nature but only a history? (Never mind whether or not this idea is tied to some misapplied Darwinian premise of "the survival of the fittest"; in his exposure of such misuse of Darwinism, Professor Flew is at his best.) This is a crucial and difficult question, sadly neglected under the historicist prejudice of our day. However that may be, for the present context it is worth noting that the so-called history of man would bear positively on questions about ethics only if what is called "history" is more than an unintelligible and aimless succession of events and epochs—in short, if the history of the race is more or less directed or tending toward some kind of a goal or end. Otherwise, "history" is merely "the past," and not a source or principle, and it has no lesson to teach except that all is flux, man included (i.e., that man is by nature an animal that *changes*, and changes himself).[6]

The second difficulty takes seriously the distinction be-

[6] In modern thought, Rousseau is perhaps the first to suggest such a view. In his *Discourse on the Origins of Inequality Among Men*, he anticipates Darwin, presenting an evolutionary account of the beginnings and development of human life and society, in which what he calls man's "perfectibility"—the power of man to change his ways to adapt to new needs—plays an important creative role. Yet how "perfectibility" means more than "alterability," in the absence of a notion or standard of *perfection* or the *perfect*, is hard to say.

tween sources and antecedents. How could it be irrelevant for our knowledge of what man is—and for ethics—if man is indeed a creature of the one God—and, what is more, is made in the image of God? Could it be irrelevant to know also whether and in what sense man stands alone in the natural world between the brutes and the divine? If the whole is governed by a Providence that is both responsible for man's being and interested in his well-being, would that not affect our understanding of *what* man is? And does not the theory of evolution by natural selection speak decisively to this question, as Professor Flew asserts? We shall return to this matter at the end.

Man's Relation to Other Animals: Uniqueness, Rank, Rule

For the purposes of ethics, it is certainly important to know man's nature but not obviously crucial to know whether he is unique. The good life for man and Martians might be the same, if they shared the same nature, even if the latter had feathers or scales rather than hair. Still, if there is something both distinctively human and also crucial to our humanity, it surely would be worth knowing, for it would probably be difficult to argue that a human being woefully deficient in such a capacity leads an admirable or enviable human life. What does Darwinism teach about man's uniqueness?

Professor Flew's argument seems sound as far as it goes. He scores the logical point that man's descent from something other than man necessarily implies his distinctness in *species*, though this does not yet show that the difference goes beyond foreclosing the possibility of interbreeding. How necessarily different, after all, are the lives and powers of two species of flies or even of mammals, related to each other as progenitor and descendant?[7] Professor Flew goes on to

---

[7] Curiously, the example Professor Flew uses to make plain the possible degree of difference between antecedent and consequent in the case of man (acorns and oaks, p. 10), if looked at closely, implies a direction and even an immanent *telos* to the process leading to man, a view that Darwinism rejects, and I suspect Professor Flew does too—though Lamarck does not. For what it is worth, I too think the matter far from closed. (See reference in footnote 12, below.)

note a few of the distinctively human capacities (p. 11), most noticeably the "unparalleled capacity for learning," and makes a proper beginning of a telling case against the re-ductionist view which denies the difference of man. I suspect he would agree that continuity of lineage does not preclude discontinuity of powers and that differences of degree may, cumulatively, give rise to differences in kind. (Here the example of the emergence, in human development, of the powers of intelligence from the unintelligent human zygote should prove the point—though some degree of intelligence is a human *potentiality* already "in" the zygote, not a mere *possibility*.)

Professor Flew, however, abstains from calling man a higher animal, thus following Darwin's maxim rather than his prac-tice. But is not man not only different and more complex but also *higher*—higher in the sense that he has the fullest range of vital powers, both in the greatest openness to the world and the greatest freedom to act in the world? Animals, unlike the heavenly bodies, can change their courses, but man alone can change his ways. And man is the only animal in the whole that is open to the whole; though only a part, man can *think* the whole. Only in man as knower does nature—including its evolution—gain that aspect of its perfection which depends on its being *known*. None of this is altered in any way by the theory of evolution: these truths were known prior to Darwin and are not destroyed by his theory.

So also was man's similarity to the animals known before Darwin. Aristotle, notwithstanding his observation that "man alone among the animals has *logos*" and notwithstanding his belief in the eternity and pure lineage of man and other species, has many passages in both the biological and ethical writings that show, on the one hand, man's animality and, on the other, the quasi-human characteristics of some animals (see espe-cially *History of Animals, VIII,* 1). That orthodox "Cartesian-ism" denies soul to animals, and sees man and beast as dis-continuously worlds apart because of man's consciousness, is an aberration that I am not certain even Descartes himself finally believed.[8] No one who has lived with dogs or horses, never mind monkeys, could make such a serious mistake. Contrary to popular opinion, and Professor Flew's, the Bible does not make it either.

The book of Genesis, and the whole Hebrew Bible, is indeed man-centered, in that it is not primarily concerned with man vis-à-vis the other animals. That should cause no difficulty, seeing that its main interest is "how *man* is to live." If squirrels or chimpanzees needed instruction on how to live, and if they had a book addressed to their well-being, we can rest assured that the main characters would be squirrels or chimps. Yet the Bible is not indifferent to animals. God shows his concern for the animals, addressing whales, the fish, and the fowl with a blessing appropriate to them (Gen. 1:22).[9] Moreover, Genesis has much to say about the animals in relation to man. Man is given dominion over the animals, but "dominion" here means in the interest of the ruled. Originally man was to have been a vegetarian, meat being allowed only later, and then reluctantly and with restrictions.[10] In Genesis 2, God makes the animals and brings them before man as possible companions, indicating their partial, albeit insufficient, kinship. During the flood, Noah is commanded to preserve *all the species,* and his uninstructed sacrifice of some of his animal roommates does not find favor with the Lord. Fi-

---

[8] See, for example, his discussion of what he calls "the teaching of nature" in the "Sixth Meditation" (*Meditations*) and in *The Passions of the Soul.*

[9] Do the literalist adherents and critics think the Bible wants us thereby to understand that these animals once upon a time understood the spoken word, even Hebrew? Is this what pre-Darwinian readers of the Bible believed? See my discussion of Darwinism and the Bible, below.

[10] Only after Noah's sacrifice reveals man's blood lust and *his* love of the smell of roast flesh (Gen. 8:20–9:4). This interpretation of Genesis 8:21 requires an argument too long to reproduce here. The reader might, however, begin by pondering how he might decide what to sacrifice or give as a gift to an unknown being in the absence of specific instructions. The permission to eat meat that is the immediate sequel is part of the covenant that addresses man's blood lust.

nally, man's own baser instincts and his "animality" are frequently asserted and demonstrated and require divinely commanded restraint (see e.g., Lev. 18: 23–24). Genesis would certainly agree with Aristotle's assertion, "For just as man, when he is perfected, is the best of animals, so too separated from law and justice he is worst of all. . . . Without virtue he is most unholy and savage, and worst in regard to sex and gluttony" (*Politics* 1253a 28–35), and also with Darwin's assertion, quoted approvingly by Professor Flew, that "man with all his noble qualities . . . with his god-like intellect . . . still bears in his bodily frame the indelible stamp of his lowly origins." (Does not Gen. 2:7 record an analogous "dual nature" to man, one part lowly, one part divine?) Man is a very special animal, but nevertheless an animal—who disagrees?

Before leaving the question of man's uniqueness, we note that Professor Flew appears to share a claim for man's uniqueness that, if taken seriously and seen for what it is, goes well beyond what the Bible grants about man's special place. Flew concludes with a favorite passage from Huxley proclaiming both man's connection with nature as well as his uniqueness. That uniqueness is said to consist in man's unexpected position as "business manager for the cosmic process of evolution," man now being able "to see himself as the sole agent of further evolutionary advance [why "advance" rather than just "change"?] on this planet." What Huxley and Flew here celebrate is the project for the mastery and possession of nature, with man as re-creator—let us be frank, as God, and not merely His image—("business manager" seems too modest a title for the "sole agent of further evolutionary advance"). The possibility and desirability of such a project was heralded by Bacon and Descartes two centuries before the theory of evolution. Darwinism can hardly be said to provide its warrant or support. The hubris of this venture—especially in the professed absence of any goals to guide the process or of any known standards by which to judge evolutionary "improvement" or "advance"—is precisely the dangerous temptation to self-sufficiency and overlordship that the Bible recognizes in man and seeks to restrain.[11]

An End for Man: Evolution and Teleology

Questions of ethics are often related to questions about the end or purpose of man. This is perhaps especially true of those who seek man's natural or proper work. The question of human ends can be raised wholly in the context of natural science (and natural theology), without appeal to revelation, in connection with the question of *teleology,* that is, the question of whether nature in general but especially living beings in particular are purposive in their doings. But in the nineteenth century, the question of teleology was considered only in relation to the question of *design,* ultimately tied to the question of God the designer. It was believed that only intelligent beings and the designed products of intelligent beings could be purposive. Understandably influenced by the powerful image of the machine, Paley saw natural order largely in terms of mechanics and therefore God as the great machinist. The examples of Aristotle and Kant suffice to show that natural order and purposive behavior need not be treated this way.

Professor Flew treats the question of natural teleology only indirectly, both in the section on Darwinian challenges to religious assumptions and in the section on the issue of progress. In the first, he argues that Darwinism challenges religious assumptions by undermining the argument to design. Overzealously pressing his overall antitheological point, he converts a hackneyed argument against Paley's *Natural Theology* (he calls Paley's, for some reason, "the *classical* statement of that most ancient argument") into an argument against the existence of God (or all gods) altogether: "Insofar as Darwinism

---

[11] A similarly dangerous naivete marks Julian Huxley's eugenic visions. For further remarks on evolution and progress, see the subsection on *Evolution and Progress,* below.

undermines the most *popularly persuasive* argument for the
existence of God, it *must* make for the complete secularization
of ethics" (p. 12; emphasis mine). (I leave it to the reader to
find his own favorite logical difficulty with this statement; the
most I think can be deduced from the protasis is that God will
not be popular, and then if and only if faith is overcome by
[this] argument.) But in the present context, we must consider
the implications of Darwin's triumph over Paley for the question of *natural* teleology.

Paley assumes that orderliness and purposiveness necessarily imply intelligent design imposed from without by an
intelligent designer, rejecting, as the only alternative considered, order brought out of chaos by chance. Darwin and Professor Flew share Paley's drawing of the alternatives: either
by art and purposive or by chance and to no purpose. Professor
Flew's comparison of natural selection with Adam Smith's invisible hand is intended to make this point: neither is directed
"by any intelligence or intention, whether divine or human"
(p. 21). If chance versus design are *in fact* the only possibilities, then natural teleology—let alone theology—is in difficult
straits.

Are these the only alternatives? Can there not be a living
process or activity that is directed intelligent*ly*—that is, in
accordance with an intelli*gible* plan—but not *by* intelligence?
Are there not orderly natural processes that mimic those in
fact due to intelligence? Are there not purposive activities that
reach an end, always or for the most part, without direction
by means of conscious planning or intention? Is not the growth
of an acorn into an oak purposive (i.e., directed to a goal)? Is
not the spider's spinning of a web or the bird's making a nest
purposive, serving preservation and procreation? Is not the
healing of a wound purposive? None of these activities involves
choice or intelligence or deliberation, yet they seem to be internally directed, natural activities regulated toward an end.
Purposiveness, or teleology, is also not refuted by the fact that
this end is served always by elaborate mechanisms that permit
the activity to continue; for purpose and mechanism are per-

fectly compatible. I have argued this point at greater length elsewhere.[12] Thus, although Darwinism *may* destroy the teleology of *imposed* design, it does not necessarily upset the teleology of *immanent* purposiveness. Indeed, a certain immanent purposiveness—that organisms are bent on survival and reproduction—is *presupposed* by and necessary to the Darwinian theory. Once this is recognized, we see that the question of natural teleology is still very much alive—though it is in fact much neglected. The *truly* interesting question is whether the Darwinians and sociobiologists are correct in referring the purpose of all organic activity to survival and reproduction or whether nature itself makes distinctions in *what* survives or even between mere life and living well. In my view, the question of the existence of a special natural end or goal for man, the attainment of which might constitute his happiness or goodness, is far from settled.

Evolution and Progress

Though much of what Professor Flew says in his section "A Guarantee of Progress?" appears sound, his exploration is heavily influenced by his suspicion that the question is of interest only to those with "secular religious cravings" (p. 18) or with "hopes of finding in living nature a passable substitute for Matthew Arnold's God" (p. 16). This seems to me a mistake. Surely any disinterested observer of the massive course of evolution would be interested to discover its tendencies. Indeed, two were identified by Darwin himself: as evolution proceeds, there are both more species and more life, that is, more individuals. (Needless to say, these tendencies are by no means *guaranteed* for the future, especially today, as Professor Flew somberly and correctly observes about the precariousness of mammalian life at least [p. 18].) A similar question can be

[12]"Teleology and Darwin's *The Origin of Species:* Beyond Chance and Necessity?" in *Organism, Medicine and Metaphysics,* ed. Stuart Spicker (Dordrecht: D. Reidel, 1978), pp. 97–120. This article also elaborates on the remarks of the rest of this paragraph.

raised about hierarchy, without any theological intent. Has there been any overall ascent (read "progress," if you prefer) in evolution? If so, what is its character? What is responsible for it? Can it be counted on to continue, barring catastrophe? Only in the end need one ask: Does this have any bearing on questions of righteousness or of ethics generally? These are all reasonable questions for anyone to ask, regardless of their moral concerns.

The recognition of progress to date is not undermined by the lack of any guarantee to further progress. Recognizing tendencies in nature *as* tendencies in nature does not require a "historicist law of development" (p. 17). I share Professor Flew's skepticism about such a law, and predictions regarding the future based on the past become increasing problematic once man enters upon the scene, for, as Professor Flew rightly observes, man's actions now affect the life of all. But can there be any doubt that the course of evolution has been in part "upward," in the sense of "higher" described above?[13] If so, what accounts for this upward tendency? Does natural selection theory give an ample account of the origin of major novelties? And what are we to make and think of the inherent potentialities of that primordial matter if, as we see, life, and then sensitive life, and then intelligent life emerged from it? Does some notion of undifferentiated, homogeneous stuff in motion, moving "of necessity" and changing "by chance" satisfy our curiosity? What is really *explained* by the suggestion that intelligence emerges out of utterly dumb stuff by a process itself utterly dumb and blind? Do we even know what we are believing when we believe this? I confess to great perplexity about all of this. I have tried elsewhere to provoke discussion of these questions by arguing and speculating in ways that go

---

[13] Bearing on this subject of ascent, see the beautiful book *Animal Forms and Patterns* by the Swiss biologist, Adolf Portmann (New York: Schocken Books 1967). See also the very provocative writings of C. H. Waddington, especially *The Ethical Animal* (Chicago: University of Chicago Press, 1967).

against the current orthodoxy. I refer the interested reader to that source.[14]

Yet the relevance of all this for ethics is, in any case, problematic at best. Even were we to find the "natural" course of evolutionary progress or ascent, even were we to know that man is the culmination of the evolutionary process, we would be a long way from a "positive evolutionary ethic" of the sort some seek and others proclaim. Still less would we be likely to possess standards to guide "further evolutionary advances."[15] Lacking such standards and such an ethic, we must stand more humbly than boldly before our powers to influence all the terrestrial globe. As there are few better teachers of human modesty than the Bible, and as Professor Flew asserts that Darwinism has dealt the Bible a mortal blow, I close with some reflections on Darwinism and the Bible.

Darwinism and the Bible

Professor Flew's treatment of this tired subject says little that is new and less that is helpful. Insofar as he is tracing the historical impact of Darwinism on popular religious belief or that of intellectuals in the Christian West in the nineteenth and twentieth centuries, he may be on the mark. But we want more than the *history* of thought: we want thought itself. We seek more than what is popular; we seek what is *so*. The inquiring mind must be prepared to go beyond all prevailing opinions in its search for what is simply true. In fact, Professor Flew does occasionally show us his own thoughts about these opinions. He does not simply describe; he endorses certain of these "historical" conclusions and exhorts us to reject others. Let us see if he counsels wisely.

[14] See note 12.
[15] For an excellent rebuttal to "evolutionary ethics," see C. S. Lewis, *The Abolition of Man* (New York: Macmillan, 1947), especially chap. 2, "The Way."

Let us make explicit and accept for present discussion Professor Flew's tacit and parochial assumption that by "religion" is meant Biblical religion and that the God of the Bible is God indeed. Is it true that "it is obviously impossible to square any evolutionary account of the origin of species with a substantially literal reading of the first chapters of Genesis"? If one in fact attends carefully and directly to those first chapters in Genesis—rather than to Blake's illustrated stories or the seemingly authoritative opinions about these chapters given by later commentators—it is not so clear, because the meaning of the text, literally read, is not so clear. For one thing, on a most literal reading, there is the notorious problem about the meaning or length of "creation days," seeing as the sun is not created until day four. Indeed, it was precisely over this question of the meaning of time that medieval controversy arose within the learned Christian community, some insisting that the text showed that creation was not simultaneous but a gradual unfolding, an *evolutio*—albeit of a prearranged divine plan.[16] Second, there is the equally well-known difficulty about the meaning of "creation." Just how and from what is God said to create? Is it creation from nothing, or is it creation by progressive separation, through the rational process of exhaustive division with regard to place and locomotion, each place filled with suitable living beings, articulated into separable kinds? Next, what is meant by "created in the image of God"? Does that mean God has human shape, or is "male and female"? Or is what is imaged in man the divine *activity* being described in Genesis 1 itself, man sharing in rationality, and, hence, in the freedom to create and the capacity to see if things are good? Are any or all of these teachings incompatible with the theory of evolution? Most fundamentally, does Darwinism force us to deny that the whole is good, and good *for us*? Or is such a judgment all the more supported when we contemplate the mysterious emergence of life, in its many-splendored forms,

---

[16] See, for example, Antonio Moreno, "Some Philosophical Considerations on Biological Evolution," *The Thomist*, 37:417–54, 1973.

and of man, who willy-nilly holds dominion, in a world not inhospitable, to say the least, to survival and well-being?

One need not go to the theory of evolution to find problems caused by the literal reading of Genesis 1–3. Indeed, the accounts of the "two creation stories" contradict each other in several details. In the first account (Gen. 1–2:3), the beginning is watery, the animals are created before man, man and woman are created together, and man is said to be created in the image of God. In the second account, the origin is dry and dusty, the animals are created after man in the effort to find him a suitable companion, woman is created later, and man has two sources, one lowly (dust) and one divine (breath of life). Perhaps the Biblical author was neither addressing nor afraid of logicians. Perhaps these difficulties, mistakenly seen as contradictions of fact by those who read these early chapters as "history," could be overcome if one could understand what their purpose was. The Biblical author does not state his intention; we have to try to figure it out from the text. Even unbelievers must assume that the book is serious (i.e., that it means to teach us important things about ourselves and the world) and, furthermore, that apparent contradictions in the story might reflect certain tensions or even contradictions in the nature of being itself (e.g., the "in-between" and morally ambiguous character of the human animal). Though this is neither the time nor the place to demonstrate it, I would encourage the close reading of the text, a reading that seems to me to reveal some plain truths about man and the world: first, about the nondivinity of the visible cosmos—in Genesis 1, only heaven and man are not said to be good—and hence also about why knowledge of the heavens (or "nature" generally) might not yield suitable or sufficient guidance for human life; second, about the failure of the inevitable attempts of human reason autonomously to determine a true knowledge of good and bad, either with or without a possibly true understanding of astronomy and cosmology.

In short, these early chapters of Genesis cast serious doubt on the adequacy of man's unaided reason to find the best way for him to live. Can Darwinsim be said to remove these doubts? Before one tries to refute the Biblical account with scientific or historical evidence, one would do well to consider whether "the truths" of Genesis are offered as science or history and, therefore, whether they are open to such refutation. Here, we should not be misled by certain Christian fundamentalists, mechanistically intoxicated nineteenth-century theologians, or imagination-poor anticlerical twentieth-century professors, whether of religion or philosophy.

On what is for Professor Flew the decisive question, the nature of man, the Hebrew Bible presents man largely as we know him, and in all likelihood as he has always been, though it never uses the term "human nature."[17] Men are shown to be lovers of beauty and of gain, foolishly overconfident in their own powers to know their own good yet dimly aware of their own puniness, devoted more to the love of their own than to the love of the good, doubters of God's existence yet filled with fear and foreboding, in pursuit of fame and dominion yet moved by awe and fear to express gratitude and to know the over-mastering power that appears to lie beyond. Some are clever, some are foolish; some are strong, some are weak; some are obedient, most are not. In speaking of men in association, the Bible shows forth the problems of founding and perpetuation, of families and cities, of law and lawlessness. It teaches the importance of a direct and didactic revelation from a higher authority—benevolent reason backed by strong force—to teach most men what they need to know and fear in order to live decently. According to the Bible itself, the preponderant char-acter of God's relation to man is as *teacher* of how to live and not as *creator* of life. Only in Isaiah does the Bible refer to the Lord as "Creator." The merely natural is subordinated to the ethical in many many ways. If one insists on the question of

---

[17] Indeed, there is no Biblical Hebrew word for "nature" or, for that matter, for "religion"; the Hebrew Bible apparently does not understand itself to have a religious as opposed to, say, a moral or political teaching. As for the idea of nature, its discovery seems to have been coincident with the begin-ning of philosophy, in the Pre-Socratic inquiry into nature.

"proofs" of God's existence, it would seem that the extraordinary character of the Biblical narrative and its teachings, not any "argument to design," is *the* most powerful evidence of the "existence of God." Indeed, it is only a lapsed ninteenth-century Christianity, burdened by a perhaps mistaken reading of its own tradition and the desire not to be caught out as unscientific, that would undertake to defend the truth or goodness of the revelation through rational arguments for God's existence, and especially with the argument to design. For Christians, as I understand it, the life, crucifixion, and resurrection of Jesus Christ is held to be true not only despite its unreasonableness, but precisely because it seems to show something beyond reason. And even for the God of Abraham, Isaac, and Jacob, if human reason were sufficient to discover Him, what need of His revelation?

If the Bible teaches the impossibility of deriving ethical knowledge—that is, knowledge indispensable for a good life—from the study of the natural world; if the Bible further denies the sufficiency of ordinary and unaided human reason to discern a *true knowledge* of good and bad (so far Darwinism should happily agree); if the Bible further shows how unlikely man is to draw the proper conclusion from these truths, that he is rather prone to what must properly be called the will to power (the *full* meaning of Professor Flew's phrase "projection of values" or Julian Huxley's "business managing" of the cosmos) and aspirations to self-sufficiency; if, in addition to these revelations, the Bible is intended to provide the remedy for all these defects; if the remedy, though not deducible, is seen to be reasonable once revealed—what are we to say about such a book and its author? Should the fact that the author does not invoke divine assistance preclude the possibility that his wisdom is more than human wisdom, perhaps even divinely inspired?

PART TWO

Marx

CHAPTER FOUR

Preliminary Thoughts for a Prolegomena to a Future Analysis of Marxism and Ethics

MICHAEL HARRINGTON

I am a democratic Marxist. I believe that Marxism is inherently democratic—indeed, that it is the quintessential democratic point of view for a collectivist age.[1] If its genuinely democratic character and implications are properly understood, I think that Marxism can make a contribution to the resolution of our society's current spiritual and ethical crisis. Today there is, as Jürgen Habermas has rightly observed, a de facto mass atheism in the West; and, like Habermas, we must wonder what, if any, secular system of ideas and values can serve, as religion traditionally has, to represent "the totality of a complex social system and to integrate its members in a unified, normative consciousness."[2] However, it would be absurd to think—as some early Marxists did—that Marxism will neatly and simply perform that function. When Marxism is degraded to such a pseudoreligion, it not only fails to perform that so-

---

[1] Cf. Michael Harrington, *The Twilight of Capitalism* (New York: Simon & Schuster, 1976), chap. 2.
[2] Jürgen Habermas, *Zur Rekonstruktion des Historischen Materialismus* (Frankfurt: Suhrkamp Verlag, 1976), p. 107.

cial–normative role but almost always rationalizes totalitarian practices.

Given this delineation of the problem, the claims I shall be making in the following pages are modest ones. The extreme tentativeness of the title of this paper underscores this point. Marxism can help us understand our spiritual and ethical crisis, and it can inform a social movement (whose character can only be intuited at this stage) that might resolve this crisis. But it is only as an informing vision underlying collective activity, not as a surrogate faith, that Marxism can enrich the ethical quality of our lives in modern society.

One further problem should be dealt with at the outset: Marxism has been misunderstood and misrepresented to such an extent that one has to anticipate prejudices and systematic ignorance when taking up the subject. There are three widely held but, I think, mistaken interpretations of the ethical implications of Marxism that ought to be mentioned. The first of these is that Marx was a hard and rather simple-minded determinist, so that for him ethical questions—questions predicated on human freedom and responsibility—do not arise.[3]

Second is the view that Marxist ethics can be reduced to the most recent ukase of a "proletarian" party. George Lukacs developed a complex argument along these lines in *History and Class Consciousness*.[4] The working class, Lukacs said, stating an unexceptionable Marxist proposition, is *the* progressive class in modern society, the historic agent of the socialist transformation. But the actual, existing workers, he continued, do not necessarily have a clear and accurate consciousness of their own class interest. Therefore, Lukacs concluded, the party is the repository of the true working-class consciousness.

Lukacs had reached this conclusion as a young man in a tortured intellectual process. But the equation of Marxist mo-

[3] Cf. Eugene Kamenka, *The Ethical Foundations of Marxism,* rev. ed. (London and Boston: Routledge & Kegan Paul, 1972).
[4] George Lukacs, *History and Class Consciousness* (Cambridge, Mass.: MIT Press, 1972).

rality with the dictates of an allegedly "proletarian"—and in fact, bureaucratic and privileged—party was to become about as simple as a decision of the Spanish Inquisition. In 1967, Lukacs looked back at that period and wrote:

> When I heard from a reliable source that Bela Kun was planning to expel me from the Party as a "Liquidator," I gave up the struggle . . . and I published a "Self-criticism." *I was indeed firmly convinced that I was right but I knew also . . . that to be expelled from the Party meant that it would no longer be possible to participate actively in the struggle against Fascism.*[5]

The Lukacs case could be probed and would add to our knowledge of Marxism and ethics, but I do not intend to do that here; the "party" version of morality is an abomination from a Marxist or any other humanist point of view and need not be taken seriously as a theory. As a practice it is, of course, of enormous significance and interest.

A third confusion has to do with Marx's intense hatred of moralizing. This is often and wrongly understood as an opposition to morality. The Germany of the 1830s and 1840s teemed with rhetorical radicals who thought that the mere enunciation of high-minded principles was a revolutionary act. "German philosophers, half philosophers and *beaux esprits*," Marx and Engels wrote in the *Communist Manifesto*,

> take over the socialist and communist literature of France which arose under the pressure of a ruling bourgeoisie and was the historic expression of the struggle against that rule. They merely forget that the emmigration of such writings from France into Germany does not bring French social conditions along with them.[6]

These "true Socialists," Marx and Engels continued, counterposed themselves to the actual democratic struggle against feudalism in the name of inapplicable French principles and thus actually helped the Prussian royal house.[7]

---

[5] Ibid., p. xxx. Emphasis added.
[6] *Marx-Engels Werke*, vol. 4 (Berlin: Dietz Verlag, 1959–). p. 485. Hereafter cited as *MEW*.
[7] Ibid., pp. 486–87.

The Marxist opposition to such merely moralizing politics should not be read as hostility to morality in politics. For example, when, a little later on in the *Manifesto*, Marx and Engels criticize their utopian socialist predecessors for seeing the workers as *merely* members of the "most suffering class," they do not intend to deny the fact of working-class suffering or to repress compassion toward it. Their point, rather, is to insist that this "most suffering class" is also—critically, decisively— a struggling, fighting class that must win its own victories rather than having them handed down to them by philanthropic radicals.

This same point is made in more complex form in Marx's economic theory. In his *Theories of Surplus Value* (often called the fourth volume of *Das Kapital*), Marx discusses the Ricardian socialists. They base themselves on Ricardo's labor theory, in which labor is the only source of value. Marx summarizes their view as follows:

Labor is the sole source of exchange value and the sole, active shaper of use value. So they say. On the other side, *capital* is everything, the worker nothing but a mere production cost of capital. But the capitalists have refuted themselves. Capital is *nothing but* thievery from the workers. *Labor is all.*[8]

The error of all this, Marx notes, is that the Ricardians "accept all the economic preconditions of capitalist production as eternal forms and will only eliminate capital, which is the basis and necessary consequence [of the system itself].[8] Against the Ricardians Marx argues that capital and profit are not thievery within the capitalist system but functional necessities of it. Again, Marx's purpose here is not to deny the moral evils of capitalism but rather to insist upon the requirements of an adequate critique that goes beyond sentimental moralizing. For him the *merely* moral condemnation of capital, which does not propose an economic system in which capital is not necessary, is as irrelevant as the revolutionary rhetoric of those

[8] *MEW*, vol. 24, pt. 3, p. 256.

German socialists who used French phrases in conditions under which they did not apply.

While this way of reading these and similar comments by Marx elsewhere in his work may help clear up some long-standing confusions, it does not fully settle the matter. Marx can be taken to task, as E. P. Thompson has correctly pointed out, not for the views on ethics so often and so wrongly imputed to him but because of his "silence" about his own ethical views. By leaving the moral dimension of his own analysis and critique implicit, Marx invited many misinterpretations.

Morality, Thompson argues, is not "some 'autonomous region' of human choice and will, arising independently of the historic process." Rather, in that historic process

> every contradiction is a conflict of value as well as a conflict of interest; . . . inside every "need," there is an affect, or want, on its way to becoming an "ought" (and vice versa); . . . every class struggle is at the same time a struggle over values; and . . . the project of Socialism is guaranteed BY NOTHING—certainly not by "Science" or by Marxism–Leninism—but can find its own guarantees only by *reason* and through an open *choice of values.* And it is here that the silence of Marx, and of most Marx*isms,* is so loud as to be deafening. It is an odd silence, to be sure, since . . . Marx, in his wrath and compassion, was a moralist in every stroke of his pen. Besieged by that triumphant moralism of Victorian capitalism, whose rhetoric concealed the actualities of exploitation and imperialism, his polemical device was to expose all moralism as a sick deceit: "the English Established Church will more readily pardon an attack on 39 of its 39 articles than on 1/39th of its income." His stance became that of an anti-moralist.[9]

But, Thompson concludes, this explicable and justifiable development led Marx to be silent with regard to the morality that breathes in his work. I hope to show that he was not quite as silent as Thompson suggests, yet I do think there is real substance in Thompson's critique.

With these three misconceptions identified, we can now proceed to the beginnings of an analysis. In broad outline, I

[9] E. P. Thompson, *The Poverty of Theory and Other Essays* (New York: Monthly Review Press, 1978) pp. 171–72.

will deal first with Marx and Engels's theories of morality, then
with some of the debates among their followers over the issue,
and finally with the relevance of this tradition under late-twen-
tieth-century conditions.

I

Karl Marx, everyone knows, held that morality is conditioned by the level of social and economic development. At
first glance that is a simple proposition. But upon closer examination the complexities of that proposition quickly emerge.

In the *Theses on Feuerbach,* Marx treats the "materialist
theory" of the relation between human conduct and social
conditions as both trite and dangerous. He is thinking of the
philanthropic reformers, like Robert Owen, who believed that
if only one were to give workers good living conditions they
would become good people. (When editing Marx's notes Engels added a reference to Owen to make the point clear.[10])
That proposition, Marx remarks critically, forgets that "the
circumstances are changed by men and that the educator himself must be educated." So this simplistic view—philanthropic
in Owen, totalitarian in Stalin, top–down and elitist in all of
its versions—is rejected. Marx counterposes "self-changing,"
or revolutionary praxis to it.[11] In short, he rejected the mechanistic theories of the relation between society and morality.

In *The Holy Family,* a sprawling polemic against the German moralizers written just a bit before the *Theses on Feuerbach,* Marx gives a sarcastic reading of Eugene Sue's novel
Les Mystères de Paris and of some of the commentaries it
inspired. In the process he defines another dimension of his

---

[10] *MEW,* vol. 3, pp. 5–6, 534.
[11] In much of the current discussion of Marxism, "praxis" is treated as the
Greek word for "practice." In fact, for Marx as for Kant, "praxis" is practice
based on a theory; that is, it is specifically human or, more precisely, it
occurs only when humans are acting in a specifically human way.

attitude toward morality and society. Fleur de Marie is a pros-
titute who, when she is first encountered in the book, feels
herself as essentially innocent, as a person who has done harm
to no one, whose "fate has not been merited." Marx approves.
Fleur de Marie has arrived at concepts of good and evil based
on her own life situation, not on ideals of the good. It is bourgeois society, he comments, which keeps her from the richness
of feeling and humanity which is in her. But when she is
converted to repentance, she "takes the filth of contemporary
society, which has touched her externally and turns it into her
innermost essence."[12] Only then does Marx criticize her.

The point here is similar to the one Marx makes much
more abstractly in the third Thesis on Feuerbach: humans are
indeed shaped by their environment, but as long as they assert
their humanity against it, they are, although degraded, good.
However, when they take blame for what is not their fault,
when they internalize social wrongs as personal sins, *that* is
a capitulation. People, even prostitutes, are never *merely* the
creatures of society. There is always some space, however
small, for the free assertion of humanity.

A little later on in *The Holy Family,* Marx develops this
theme in a way that will bear very much on one of his most
mature and important discussions of morality and society, the
Critique of the Gotha Program (which will be discussed below). The law, he argues, is always one-sided; it always abstracts from the specific humanity of the criminal and treats
the criminal under a general rule. Under more humane (and
human) conditions, Marx comments, punishment would be
the judgment a person passes on himself. It would not come
from the outside, as an abstraction imposed upon him by state
power. The idea of external and compulsory punishment vi-
olates the basic humanity of the individual. Here again, what
is human is what is self-determined.[13] In 1853 Marx returned

12 *MEW*, vol. 2, pp. 180–85.
13 *MEW*, vol. 2, p. 190.

to this very same theme, in a biting attack on the death penalty, where he condemns Kant's theory of punishment as "only a metaphysical expression of the old 'Jus Talionis.'"[14]

Marx's attitude toward capital punishment points in the direction of another, much larger theme, one which has been the *locus classicus* of some of the most influential misreadings of the Marxist theory of ethics. It has to do with the relationship between necessity, or inevitability, and moral rightness.

When Marx attacked the death penalty, he assumed its basis in capitalist society. Both he and Engels believed that capital punishment and a criminal system based on the lex talionis were *inevitable* in a class-dominated society. For Marx, only when the social order was truly human would it be possible for individuals to be human. But Marx understood that, in a capitalist society, his own or any other moral objection to capital punishment had to be based upon an ideal that does not yet exist. The difference between him and the moralizers is not that they posit ideal values while he merely accepts reality. It is rather that Marx's ideals are related to social movements that can bring them about, while his academic friends substitute their ideals for those social movements. For Marx, what is historically inevitable—and even what is progressive— is not necessarily good and moral.

A misunderstanding of Engels's famous maxim that freedom is the recognition of necessity has led generations of "Marxists" to turn Marx upside down on this count. The man who believed that Fleur de Marie was virtuous precisely to the degree that she railed against the inevitabilities afflicting her— and that she was corrupted precisely in accepting those inevitabilities as her own fault—has been turned into the proponent of a vulgar pragmatism in matters of morality.

Franz Mehring, the great Marxist historian and left socialist, understood the fallacy of this misinterpretation of Marx quite well. Writing in the context of the debate over neo-Kantianism that took place within the German Social Democratic

---

[14] *MEW*, vol. 8, p. 508.

party, he observed that "the ethical *evaluation (Beurteilung)* of a social event is something quite different from a genetic *explanation* of its development. Marx in no way admired capitalism because he demonstrated its historic necessity."[15]

This distinction between an ethical evaluation (or justification) and a historical explanation is critical if one is to understand *The Communist Manifesto*. In that document, Marx advocated the eventual overthrow of capitalism at a time when he believed that Marxists had to work, in the here and now, with the left wing of the bourgeoisie. That tactical adaptation recognized that there are (unfortunate) limits on the ideals that can be brought to life in a given historic moment. But the ideals are not defined by the limitations; they are, so to speak, asserted in spite of the limitations.

Thus we need to look more closely at Engels's comments on freedom and necessity. "Hegel," he wrote in *Anti-Dühring*,

was the first who correctly presented the relationship between freedom and necessity. "Necessity is blind *only to the degree that it is not conceptually grasped.*" Freedom does not consist in the celebrated independence from natural laws, but in the knowledge of these laws and in the consequent possibility to let them work, in a planned way, to specified ends.[16]

There is certainly some ambiguity in this formulation, but it is obvious that the reference to planning and the specification

---

[15] H. J. Sandkuhler and R. de La Vega, eds., *Marxismus und Ethik* (Frankfurt: Suhrkamp, 1974), p. 363. Hegel made this same point in a book that Marx studied quite carefully and commented on at great length, the *Grundlinien der Philosophie des Rechts* (Hamburg: Felix Meiner Verlag, 1955). He wrote that "an analysis on the basis of historic grounds should not be confused with a philosophical analysis based on the concept, and historic explanations and justifications should not be equated with justifications valid in and of themselves" (p. 23). Roman family law, he said, was clearly a product of Roman society—on this count, as on so many others, Hegel was a "Marxist," or Marx was a Hegelian—but that by no means proves that it was reasonable or right.

[16] *MEW*, vol. 20, p. 106. *Anti-Dühring* was a polemical and propagandist book that contains more than a few other simplifications.

of goals means that Engels envisions humans freely choosing ends and using natural laws to forward them. One recognizes necessity precisely in order to be freed from its blind consequences and to be able to utilize it for human purposes.

Now as I said, Engels has often been taken to have maintained that what is necessary is moral (i.e., that the liquidation of 6 million peasants was an unfortunate but progressive moment in the extending of man's control over nature). Even if this was Engels's meaning, did Marx hold the same view? There is much evidence that he did not. In his early critique of Hegel, for example, the young Marx brilliantly denounced just such a theory. Hegel, the great antipositivist, was, Marx argued, actually a positivist in that he simply accepted reality— reconciling humanity to necessity rather than calling for a change in reality.[17] Still, it could be objected that this was merely the critique of the young Marx who had not yet passed over that *coupure* (in Althusser's vocabulary), that intellectual gulf, and become the mature, neo-Stalinist Marx.

But to see that this was not merely some premature idealism, consider Marx's comment on the British in India in 1853. This—like his article against capital punishment—was written eight years after the putative *coupure* (i.e., it is an analysis by the mature Marx). Moreover, it presents an ethical analysis in the context of a concrete political case. In his August 8, 1853, dispatch to the New York Tribune, Marx was positively enthusiastic about the progressive consequences that would follow from British imperialism in India.[18] He quite inaccurately held that what we would now call the "spread effects" of imperialism would transform the entire society and lay the basis for a socialist movement. In short, British imperialism was a progressive necessity in India. But did that mean that British imperialism was moral? Or that it was deserving of political support? That was precisely the inference drawn by the proimperialist socialists of the Second Interna-

---

[17] *MEW*, Erg. Bd., pt. 1, p. 581.
[18] Shlomo Avineri, ed., *Karl Marx on Colonialism and Modernization* (New York: Doubleday/Anchor, 1969), pp. 132ff.

tional prior to World War I. And it is precisely the inference that Marx refused:

All the English bourgeoisie may be forced to do will neither emancipate nor materially mend the social condition of the mass of the people, depending not only on the development of the productive powers, but of their appropriation by the people. *But what they will not fail to do is to lay down the material premises for both.* Has the bourgeoisie ever done more? *Has it ever effected a progress without dragging individuals and peoples through blood and dirt, through misery and degradation?* The Indians will not reap the fruits of the new elements of society scattered among them by the British bourgeoisie till in Great Britain itself the now ruling classes shall have been supplanted by the industrial proletariat, or till the Hindoos themselves shall have grown strong enough to throw off the English yoke altogether. . . . *The profound hypocrisy and inherent barbarism of bourgeois civilization lies unveiled before our eyes, turning from its home, where it assumes respectable forms, to the colonies, where it goes naked.*[19]

So—for the "mature" Marx British imperialism is progressive; it drags individuals and peoples through blood and dirt, misery and degradation; it is profoundly hypocritical and inherently barbaric; and the political attitude to adopt toward such a phenomenon is to overthrow this agent of "progress" at the earliest possible moment. It would be hard to imagine a clearer example of the difference between a genetic explanation (that "objectively," British imperialism creates the material premises for Indian emancipation) and an ethical judgment (that British imperialism is barbaric).

Finally, some of these soaring themes might be personalized. Marx, it is well known, led a difficult life and for years had to count every penny. Yet he did not ordinarily complain of his lot. In 1867, however, he wrote to Sigfried Meyer in New York and during a particular trying time said,

Why haven't I answered you? Because I have been living on the edge of the tomb. So I must use *every* moment in which I can work in

[19] Ibid., p. 137. Emphasis added.

order to finish my book and I have sacrificed everything, my health,
my happiness, my family, to that end. . . . If one wants to be an ox,
he can simply turn his back on the sufferings of human kind and
look after his own house.[20]

Thus as he was completing volume I of *Das Kapital*, he made
it quite clear to Meyer that he regarded his own life as a moral
choice.

In both his theory and his own life, then, Marx believed
that society sets limits upon choice but that within those limits,
men and women *should* always represent the claims of human
self-determination against the necessities—natural, social, or
political—of external domination.

Friedrich Engels, a lesser genius (but, let it never be forgotten, a genius in his own right), was given the polemical
assignment in the division of labor which he and Marx worked
out. His simplifications are treasured by those who want their
Marxism desiccated and scientistic. Yet even when he did put
things too neatly, Engels was more complex than most of his
admirers—and much more morally admirable. There are, he
wrote in *Anti-Dühring*, three principal contending moralities
in European society, each linked to a social point of view. These
are the Christian–feudal morality, the bourgeois morality, and
the proletarian morality. "Which is true?" Engels asked. "No
one of them in the sense of absolute finality. But obviously the
one which has the most long-lived elements is the one which
represents the future in the present, the transformation of the
present. Thus it is the proletarian morality."[21]

Note that, even in this propagandistic and polemical book,
Engels does not claim that proletarian morality is finished and
final. And a few paragraphs later he stresses that all three
moralities necessarily have "much in common" because they
share a common background and function in the same society.
However, Engels does insist that "men's concepts and ethical
views [*Anschauungen*] are *in the last instance* derived from

[20] *MEW*, vol. 31, p. 542.
[21] *MEW*, vol. 20, p. 87.

[*schöpfen aus*] the practical relations in which their class situation is based—out of the economic relations in which they
produce and exchange."[22]

In *Anti-Dühring* Engels was already employing the phrase
"in the last instance," which he was later to use when he tried
to exorcise some of the determinist readings which, he freely
admitted, he and Marx had sometimes invited. I do not want
to get into what the phrase means, since there is a small library
of debate on the issue. For the present purpose, I merely want
to stress that Engels included a considerable amount of in-
determinacy in his theory of the ultimate social origins of moral
theories and values. Indeed, again in *Anti-Dühring*, Engels
himself underlines this point. The moral and legal points of
view, he comments "are more or less corresponding expressions—positive or negative, ratifying or countering—of societal
and political relations."[23] So the same economic and social
relations can give rise to *counterposed* moralities.

So there are no simple guidelines, not even for historical
explanations of moral phenomena, much less for normative
judgments about the goodness or badness of specific individuals or social acts. And finally, just to complicate matters a bit
more, Engels posits a non-class morality as the ideal, "that
truly human morality which first becomes possible at that
societal stage when class antagonisms have not only been
overcome [in communism] but have been forgotten in the
praxis of life."[24] The class character of morality, then, is a
limitation; the most fundamental value has to do with a hu-

---

[22] Ibid. Emphasis added.

[23] Ibid., p. 89. During the debate between the orthodox and neo-Kantian Marxists, Max Adler, the Austro-Marxist (and one of the most brilliant of the
neo-Kantians), attacked Kautsky's Darwinist version of the Marxian ethic.
"The very same material basis," Adler wrote, "can make very different social
orders possible. If there were not a [socialist] ethical ideal, why shouldn't
the proletariat ultimately settle for a system of industrial feudalism in which—
and this is not to be excluded as a possibility—it would receive better wages
than now, clean housing, shorter working time and sufficient insurance
against sickness, accidents, old age and being invalid." [*Marxistische Probleme* (Berlin–Bonn: Dietz, Verlag, 1974), p. 134.]

[24] *MEW*, vol. 20, p. 88.

manity that can only emerge in the classless society. This points us toward a critical theme: the Marxist theory of human nature.

Have I not just turned Karl Marx into Ludwig Feuerbach? "Feuerbach," Marx wrote in the *Theses:*

resolves the religious essence into the *human* essence. But the human essence is not an abstraction which dwells in the individual. In its reality it is the totality [ensemble] of the societal relations.

Feuerbach, who does not go into the critique of this real essence, is therefore forced (1) to abstract from historic process, to make of the religious temper something *per se,* to presuppose an abstract— *isolated*—human individual; (2) to conceptualize the essence therefore only as "genus" [*Gattung*] as an inner, mute universal characteristic which binds individuals together through *nature.*[25]

This is, however, *not* a rejection of the idea of human nature but rather a rejection of one version of that idea. In many religious conceptualizations—obviously including the Judaeo–Christian—human nature is fashioned by an act of God. In sophisticated readings of Genesis that account of creation can be, and has been, made quite compatible with evolutionary theory. Even so, it is, I think, necessary for those interpretations to insist upon the divine role in the creation of humanity, even if all of the secondary causation is taken naturalistically. Feuerbach was an atheist and opponent of religion yet, Marx is arguing here, he takes over that traditional, religion-based concept. As a result, he abstracts humanity from history and does not understand that human nature is a social (collective, conscious, historic, and not merely natural) self-creation.

I agree with Marx on all these counts and, like him, adhere to a historical, transformative concept of human nature. That concept, in one sense, is not at all original with Marx. It is to be found in much of classic German philosophy—in Kant, Fichte, Hegel, Schiller, and others. Marx's youthful statement

[25] *MEW*, vol. 3, p. 6.

of the theme in the Paris manuscripts shows how deeply he drew from that tradition:

> free conscious activity is the species character of men . . . The animal is immediately one with its life activity. It does not differentiate itself from those activities. It is *them*. Man makes his life activity into an object of his will and consciousness. He is not something determined which is immediately one with the determination. Conscious life activity immediately differentiates human activity from animal activity. Precisely in this way is man a species being [*Gattungswesen;* a complex term which here can be read as "social being"]. Or he is only a conscious being, i.e. his own life is for him an object, precisely because he is a species being. Only in that way is his activity free activity.[26]

But cannot these formulations be dismissed as the vaporings of a young philosopher who had not yet recovered from a bout of Hegelianism? Not at all. One finds this theme running through Marx's writings throughout his life. That definition of human nature is, for instance, repeated in all of its essentials in the third section of *Das Kapital* at the beginning of the chapter on the labor process.[27] Moreover, the young Marx emphasized his—and the proletariat's—debt to his philosophic forerunners when he said that the coming revolution would "effectuate" philosophy in general and specifically that philosophy for which humanity was of the essence.[28] Many of the values which he sought had just been worked out by bourgeois thinkers who went before him.

Now it is true that Marx gave up some of the romantic formulations of that specific period. In particular, his contact with the real working class led him to revise some of his bookish abstractions about the proletariat. And yet it is also true that throughout his life he took as axiomatic the values of freedom and humanity, which were in no way original with him. The importance of this point can be illustrated by con-

[26] *MEW*, Erg. Bd. 1, p. 516. Emphasis added.
[27] *MEW*, vol. 23, pp. 192–93.
[28] *MEW*, vol. 1, p. 391.

sidering an incident that occurred during the neo-Kantian debate within the German Social Democratic party alluded to earlier.

Some of the neo-Kantians had argued that Kant's dictum that one must never treat human beings as means was the statement of the basic socialist ethic. There were some in those discussions—Rudolf Hilferding, for instance—who thought of Marx as "merely" a scientist and argued that it would be possible for a bourgeois to accept Marxism as a scientific prediction of an inevitability which he would fight against. For some of those who thought along those lines it was necessary to "complement" Marx with a Kantian ethic.

Against this view Mehring argued:

> If we look at the Communist Manifesto again, we read there that Marx demanded, as the ideal situation for human society, an association in which the free development of each is the condition for the free development of all. In terms of content, this is about the same as the key proposition of the Kantian ethic upon which the neo-Kantians lean so heavily: "Act so that you treat humanity, in your person as well as in that of everyone else, always as an end, never merely as a means." In content, ethics in Marx and Kant are thus the same. Only in Kant's "analytic grounding," he knew how to unite this proposition with the medieval, estate separation of citizen [*Staatsbürger*] and worker [*Staatsgenossen*]. In Marx's "historic–causal" grounding, he understands out of economic development how to effectuate his ideal.[29]

It is not, in short, the formulation of the ideal which is distinctly Marxist but the social, historic, economic—and revolutionary political—understanding of how the ideal could become practical.

What is involved here has both historical and present relevance. In the *Phenomenology*, Hegel rightly criticized Kant's

---

[29] Sandkuhler and de La Vega, eds., *Marxismus und Ethik*, p. 365. The difference between *Staatsbürger* and *Staatsgenossen* is much more complex than the difference between citizen and workers, which is my translation. In this case, I think it adequate. Mehring refers to the fact that Kant did not believe that workers should have the right to vote, among many other things.

formalism and pointed out that his ethical maxims were compatible with counterposed responses to actual cases. In one sense—that of the highest generality—Karl Marx is profoundly a product of Western philosophy and takes over many of its values. In this guise, what is important about him as an ethical thinker is the politics of his morality, not his morality "as such." (There is a major exception to this statement—it is found in the *Critique of the Gotha Program*—which will be dealt with shortly.) Second, Hegel identified an important element in the present ethical crisis. With the subversion of religion-based ethics through the de facto triumph of agnosticism and atheism, all of the handbooks of religious–ethical rules have been utterly devalued. Everyone, including the more astute theologians, has become existential, situational, and the like. But that provides no basis for solidarity, for a common, consensual normative consciousness, as Habermas put it. That problem will be faced in the third section of this paper.

Now however, let me return to the main line of argument. If Marx believed that the expansion of human freedom was the highest value, how does that express itself in his ethical theory? Leon Trotsky and John Dewey agreed on how to answer this question—I am not sure that knew they were agreeing, but that seems to me to be the fact—even though Dewey was not, of course, any kind of Marxist. Trotsky wrote:

From the Marxist point of view, which expresses the historical interests of the proletariat, the end is justified if it leads to increasing the power of humanity over nature and to the abolition of the power of one person over another. "We are to understand that in achieving this end anything is permissible?" demands the philistine sarcastically, demonstrating that he understood nothing. That is permissible, we answer, which *really* leads to the liberation of humanity. . . . When we say that the end justifies the means, then for us the conclusion follows that the great revolutionary end spurns those base means and ways which set one part of the working class against other parts.[30]

---

[30] Leon Trotsky, John Dewey, and George Novack, *Their Morals and Ours* (New York: Pathfinder Press, 1975), pp. 48–49.

In his reply, Dewey began by quoting Trotsky:

"That is permissible, we answer, which really leads to the liberation of mankind." Were the latter statement consistently adhered to and followed through it would be consistent with the sound principle of the interdependence of means and ends. Being in accord with it, it would lead to a scrupulous examination of the means that are used, to ascertain what their actual objective consequences will be so far as is humanly possible to tell—to show that they do "really" lead to the liberation of mankind.[31]

Dewey believed that Trotsky did not do this, that he had, on completely *a priori* grounds, chosen some means and discarded others. Dewey was, I think, both faithful and unfair to Trotsky on that count—but that does not concern me here. I take the principle that both Trotsky and Dewey accepted as valid—and as telling one very little about what to do in a given siuation.

Thus far I have addressed the ethical values Marx shared with non-Marxists, like Dewey and Kant. Does that mean that the distinctive Marxist contribution to ethics is confined only to the politics of morality, to his strategy for effectuating the norms established by the German classical philosophy? It does not.

The most striking case of a specifically ethical discussion in Marx's work is to be found in his "gloss" on the program of the German Workers' party, adopted at a congress in 1875. This *Critique of the Gotha Program*, as it is usually called, was originally a private communication from Marx to Wilhelm Bracke, meant to be shared with the Marxist leadership of the German movement. It was published in 1891 by Engels and has been rightly considered a classic statement of Marx's view. It should come as no surprise to those familiar with the enormous influence of misinterpretations of Marx that this document is regularly cited as a defense of policies which are the

---

[31] Ibid., p. 69.

ethical antithesis of everything for which Marx stood. That tragic irony will surface a little later on.

The Gotha Program had called for a "more just distribution of the product of labor." Marx comments on this, echoing his earlier critique of the Ricardian socialists:

Doesn't the bourgeoisie claim that the present distribution is "just"? And isn't it in fact the only "just" distribution on the basis of the present mode of production? Are economic relations ruled by concepts of justice [*Rechtsbegriffe*] or isn't it the case that the relations of justice on the contrary derive from the economic relations? Haven't the socialist sectarians the most varied images of a "just" distribution?[32]

Here Marx means to underscore the specificity of a "capitalist" form of justice. The possibilities of social justice in a capitalist system are limited by the structure of that system, but they are not determined by its structure since, as Marx was well aware, within the systemic parameters of capitalism different patterns of distribution—a greater or a lesser in equality—can be found. The extent to which the full, albeit limited, potentialities of "capitalist justice" would be realized in any particular society was, he argued, a function of the class struggle within the society. Moreover—and here again Marx is attacking the moralizers, not morality—if one leaves conceptions of justice to the imagination of sectarians, there will be no agreement. Justice and morality must be conceptualized, like everything else, historically, in relationship to a given society. This point then leads Marx to make a fateful distinction, one that has been twisted to profoundly anti-Marxist purposes in the Soviet Union and other "communist" countries.

Socialism in the full sense of the word, Marx is going to argue, is not based on equality for all. But before he comes to that critically important point he distinguishes two stages in the socialist transformation: the transitional, or lower, stage of communist society and fully developed communism. This is

[32] *MEW,* vol. 19, p. 18.

sometimes presented as the distinction between socialism and
communism. In the transitional period, Marx writes,

> we deal with a communist society, not as it has *developed* itself on
> the basis of its own principles, but, on the contrary, as it *emerges* out
> of capitalist society and thus still burdened in every one of its rela-
> tions—economically, morally, spiritually—with the afterbirth of the
> old society out of whose womb it came. Consequently, the individual
> producer receives—after the deductions [of the funds for social con-
> sumption]—exactly what he gives to the society.[33]

This two-stage theory of communism was put to work by
Joseph Stalin and his comrades to justify the most outrageous
violations of fundamental Marxist principles. If asked why
workers had no right to organize politically or economically,
why inequality not only persisted but was inceasing in the
Soviet Union, and so on, the official answer was that in the
period of transition such things happened but they would all
cease with the arrival of full communism. Moreover, they are
ordered from *above*, by a state standing above the workers.
And the date at which the second, full stage of communism
is to begin keeps getting pushed into the far distance. When
Nikita Khrushchev, in a fit of exuberance, announced that the
second stage was at hand, he helped prepare the way for his
own removal, since his heresy would have required—even in
Stalinist terms—the reduction of the coercive apparatus of the
Soviet state. But the actual fact of the matter is that the Soviet
practice in this area is light years distant from anything envisioned by Marx and Engels. For them, as Herbert Marcuse
remarked in *Soviet Marxism,* "there would be no coercive state
organs separate from and above the associated laborers, for
they *are* the socialist state."[34]

The *Critique of the Gotha Program,* a document which,
as we will see in a moment, is one of the most libertarian

---

[33] *MEW*, vol. 19, p. 20.
[34] Herbert Marcuse, *Soviet Marxism* (New York: Columbia University Press,
1958), p. 21. Part II of this book is devoted to an interesting discussion of
Soviet ethics.

ethical statements in all of Marx's writings, has historically been used to justify the totalitarian repression of working-class freedom. Only those who are shocked by the "cunning of reason" in history have the right to be surprised by this turn of events.

Almost immediately after the discussion of the stages of socialism Marx develops what I have just called his libertarian ethical position. The individual producer in his example, Marx comments, is paid according to his or her work and as a result "equal work in one form is exchanged against equal work in another form." He then writes:

Equal right is thus still—in principle—*bourgeois* right, though principle and praxis are no longer at odds which they are when the exchange of equivalents in commodity exchange only occurs *on the average* and not in the individual case.

Despite this advance, this *equal right* is still restricted under bourgeois limits. The right of the producers is *proportional* to their output; the equality derives from the fact that the output is measured by *an equal standard,* by labor. One person is physically or spiritually superior to the other and can do more in the same period of time or work longer; and labor, in order to serve as a measure, must be determined according to its duration or intensity, or else it would cease to be a measure. *Equal* right is unequal right for unequal labor. It acknowledges no class differentiation, because each worker is like every other worker. But it silently acknowledges unequal individual gifts and the consequent unequal productivity [*Leistungsfähigkeit*] of the worker as natural privileges. *It is therefore, like all rights, a right of inequality in terms of its content.* Rights can, by their very nature, only arise out of the application of an equal measure; but unequal individuals (and they would not be different individuals if they were not unequal) are only measurable by an equal measure so long as one brings them under a single perspective, observes them from only one *specific* side, e.g. only as workers, and sees nothing else in them, abstracts from everything else. . . . But these grievances are unavoidable in the first phase of communist society as it emerges out of the long labor pains of capitalist society. Right can never be higher than the economic structure (*Gestaltung*) and the cultural development of society which that structure conditions.

In a higher phase of communist society, after the slavish subordination of the individual under the division of labor and the conflict between spiritual and bodily labor vanishes; after labor is no longer a means to life but the most basic need of life; after the all-sided development of the individual and his or her productive power and all the springs of cooperative wealth flow—then, for the first time, will it be possible to stride beyond the narrow horizon of bourgeois rights and for society to inscribe on its banner: from each according to his or her capacity; to each according to his or her needs.[35]

Let me stress only two aspects of this extraordinary passage. First, the objection made to the abstract character of law, which must always prescind from the individual quality, is identical with the critique made by Marx in the *Holy Family* thirty years earlier. Second, and much more substantively, it would be hard to provide a clearer illustration of Marx's personalism, his deep commitment to the freedom of the human being who, in Marx's ethical ideal, chooses his or her life in the fullest sense of the term and, even though enabled to do so because of the social development and control of the economy, is thereby given the right to the fullest individualism.[36]

In summary, then, Karl Marx's deepest values are based upon a vision of a historically developing, self-creating human nature, that progressively reveals the capacity for life in a society of social and noninvidious individuality. In the here and now, one judges individuals in terms of how they assert that humanity to the degree possible in their particular situation. And one judges social movements not simply in terms of their economic "progressiveness" but most decisively in terms of how they forward or hinder the growth of human autonomy. In the case of both individuals and movements, there is social-class determination in that neither is a tabula rasa and in that both must be seen within the context of a social economic formation they have inherited; but neither is totally determined

[35] *MEW*, vol. 19, pp. 20–21. In German, adjectives and pronouns take the gender of the noun. My translation of "his or her" is thus my own choice.
[36] Marx was not an "individualist" in the sense which that term has in Western culture and philosophy. "Personalist," as Mounier used the word, more aptly describes his views. In any case, I am impressed by the limits and poverty of language at this point. The right word does not yet exist because the right reality does not either. If the utopia envisioned by Marx ever comes, its vocabulary, not just its economic and social life, will be richer than ours.

and each is to be judged on the basis of how well it utilizes whatever freedom is available to it.

This view, as I have stressed at a number of points, has been profoundly and normally misinterpreted. Therefore, in the second section of this paper I will look at two of the controversies over Marxism and ethics in order to communicate some sense of the arguments that developed in this area after Marx's death. We proceed from Marx and ethics to Marxism and ethics.

II

There was a debate over Marxism and ethics in the European socialist movement between 1890 and the Great Depression. Its roots are to be found primarily in the period of relative capitalist growth and stability between 1890 and 1914.

On the one side, there were socialists who said that the movement had to go "back to Kant," that Marxism was without an ethic and had to borrow one from the great German philosopher whose philosophy had enjoyed a revival from the 1860s on. The back-to-Kant movement was, with some notable exceptions, on the right wing of German socialism—that is, it sympathized politically with Edward Bernstein and wanted theories corresponding to the actual, evolutionary practice of the Social Democratic party and the unions. These revisionists were much more realistic about the character of their times than were their orthodox Marxist opponents—but they did not know how transitory those times were. Orthodoxy was defended by Karl Kautsky, who based himself on a natural–scientific reading of the master's works.

The period abounded in paradoxes. On the one hand, the obvious was true: that socialist eschatology had lost at least some of its lure in a prospersous Germany and that this fact provided a base for revisionism.[37] But the counterintuitive was

---

[37] Jose Luis L. Arangueren, *El Marxismo como moral* (Madrid: Allianza Editorial, 1970), p. 112.

also true: As Karl Korsch pointed out, the Marxists made little headway during the recessions and persecutions of the 1870s and 1880s and considerable gains in the calm and expansiveness of the 1890s.[38] Here we enter a tangled and complex history.

Following Engels's dictum at Marx's funeral—that "as Darwin discovered the laws of organic nature, so Marx discovered the developmental laws of human history"[39]—the orthodox Marxists increasingly interpreted their master in a positivistic, scientistic fashion. One of the most extreme statements of this view was made in 1910 by Rudolf Hilferding in his extremely important book *Das Finanz-Kapital*. Marxism, Hilferding wrote, is "value-free," and, he continued:

It is therefore a false concept, even though a widely shared false concept both within and without the movement, to simply equate Marxism and socialism. For logically, abstracted from its historic impact, Marxism is only a theory of the laws of motion of society, which the Marxist theory of history formulates in general and Marxist economics applies to the epoch of commodity production. The socialist consequences are the results of tendencies which develop in the commodity-producing society. But the insight into the rightness of Marxism, which includes the insight into the necessity of socialism, is in no way the product of value judgments and just as little a direction for practical affairs. It is one thing to recognize a necessity and another to put one's self in the service of this necessity.[40]

Two years later Franz Mehring made the same point in a particularly dramatic way. In an essay on Fichte, Mehring attacked

all the attempts to orient the proletarian class struggle on the basis of philosophic concepts and conceptual games (*Begriffsspielereien*). They must be swept away from the door like spider webs which can only conceal from the workers the necessary knowledge of nature and history. In the heavens of our intellectual world, the German

---

[38] *Marxismus und Philosophie* (Vienna: Europäische Verlagsanstalt, 1966), p. 43.
[39] *MEW*, vol. 19, p. 335.
[40] *Das Finanz-Kapital*, vols. 45–46 (Berlin: Dietz Verlag, 1947).

classic philosophy is a distant star whose dazzling rays can delight, or blind, our eyes but can neither warm our hearts nor tighten our muscles. Our confidence ripens in the sun of labor and the least proletarian, who puts all his strength in the economic and political struggle for the emancipation of his class, takes the road to human freedom more surely than the great thinkers of our classic philosophy who groped in dense clouds and ultimately could not find that way to freedom.[41]

This is not to suggest that Mehring was insensitive to ethical values. In 1900 he had replied to one of the neo-Kantians, Hermann Cohen of Marburg. Cohen, as Mehring summarized his view, insisted that

socialists shall not give up the idea of God which needs to be understood as . . . nothing but the belief in the power of the good, the hope in the coming of justice. Secondly, justice and state are ideas to which socialism must render homage since, just as there can be no freedom without law, so can there be no free personality, no true community of moral beings without a community dedicated to law.

Mehring answered that Cohen's points (I have omitted a third which is not relevant here) had already been taken over by the socialists.

Does Herr Cohen really believe that the social democratic party has gone through all the battles of the last forty years, and particularly [illegalization under] the anti-socialist laws for twelve years, without a belief in the power of good and confidence that justice would triumph?[42]

But if Mehring recognized the reality of ethical motivations in the struggle, his version of Marxism provided no grounding for them. How, given this view, could one provide a basis for action? If Marxism is a science and not even necessarily connected to the socialist struggle, if the values of the

[41] *Zur Geschichte der Philosophie* (Berlin: Soziologische Anstalt, 1931), pp. 89–90.
[42] *Marxismus und Ethik,* p. 361.

German classical philosophy are irrelevant and even inferior
to the instincts of the least proletarian, where is there a place
for morality?

In 1922 Karl Kautsky tried to answer these questions in
his book *Ethics and the Materialist Conception of History*.
The animal world, Kautsky argued, develops a social drive in
order to help its members in the struggle for existence. A
morality develops out of this functional coming together (i.e.,
rules for the herd). But animals do not, Kautsky emphasized,
develop ethical ideas. "Only humans are capable of posing
ideals and striving for them."[43] What, then, is the basis of
those ideals? It is not to be found in human nature or the
eternal dictates of reason, as the atheists and materialists of
the eighteenth century thought in company with the theists
and idealists. Darwin, Kautsky holds, put an end to this thesis
once and for all. Yet even before Darwin there was a theory
that "also" revealed the secret of the ethical ideal—the theory
of Marx and Engels.

What differentiates men from beasts, Kautsky continued,
is not the production of the means of consumption—there are
animals which do that—but the production of means of production, of tools. At that point "begins the human becoming
of the anthropoids [*Tiermenschen*]."[44] But this human becoming is not simply to be equated with moral progress. "One can
even say that the ape is more human, more man-like than
man. The murder and slaughter of members of one's own
species for economic reasons are a product of culture, of the
arms technology."[45] The ethical ideals, however, arise out of
the practical needs of a mode of production, a fact that is
unknown to most of the people who follow the rules and which
is only accessible to the scientific investigator. For modes of
production are not simply characterized by technological development but "by a specific intellectual content and capacity,

---

[43] *Ethik und materialistische Geschichtsauffassung* (Bonn: Dietz Verlag, 1973),
     p. 68.
[44] Ibid., p. 80.
[45] Ibid., p. 99.

a specific conception of the process of cause and effect, a specific logic, in short, a specific mode of thought."[46]

Finally, Kautsky asserted that, although the proletariat needs an ideal, it cannot find this ideal in *scientific* socialism, "the scientific investigation of the developmental and transformational laws of the societal organism for the purpose of establishing the *necessary* tendency and goal of the proletarian class struggle." The socialist investigator is also a fighter in the cause. "So, for instance, in a Marx the effect of his ethical ideals as well as his scientific work break through. But he was always concerned to banish the ethical from his work as far as he can, and rightly so."[47]

Finally, we should consider the position taken by one of Kautsky's principal antagonists in this debate, Max Adler. Adler was a neo-Kantian *and* a Marxist. He was one of the very first to recover—or perhaps he never lost—a sense of the importance of the "Hegelian" emphasis on will, values, ideas in Marxism and of Hegel's own influence in the development of Marx's thought. But what concerns us most here is Adler's critique of Kautsky's views on ethics.

Initially, Adler congratulates Kautsky on his rejection of all relativist theories of morality.[48] That, as I will emphasize in section III of this paper, is of great contemporary importance, for the problem today is precisely one of finding a secular and rational basis for a consensual normative consciousness in an atomized, hedonistic, and individualistic age. What Adler and Kautsky stress is precisely the emergence of something enduring in the historic flux. That "something," I argue, is the unfolding of the human potential, which is not a providential occurrence as in Hegel but a historic possibility which is still open.

But if Adler agrees with Kautsky that one must seek a rational, objective basis for values, he profoundly disagrees on

---

[46] Ibid., p. 123.
[47] Ibid., p. 141.
[48] *Marxistische Probleme*, p. 107.

the way that Kautsky carries out this task. Above all, Adler points out that a genetic, evolutionary, historic account of *how* values develop in fact is not a description of what the moral sense *is*.[49] As a Marxist, Adler agrees with Kautsky that values have no place in a scientific analysis of the interconnections and causalities of history. But as a neo-Kantian, he rejects the notion that scientific analysis is sufficient to provide an ethical basis for action. Marxism, he argues with considerable subtlety, can and must confront the ethical as a historic force; but only Kantianism can provide the missing justification for values.

I disagree with both Kautsky and Adler. Both thinkers were mistaken in believing that the fact–value distinction can be maintained in the social sciences. It is of the very essence of Marxism to argue that every fact is seen from a vantage point—a moment in history, a position in the class structure, which affects even the most idiosyncratic observer—and that among these various perspectives, that of the proletariat is more valid than any other. There is, Marx holds, a positive, ontological value to the solidarity that develops within the working class. If one must then add, as we will see in a moment, that many of Marx's expectations about that solidarity have been disappointed, then we must face up to the fact that—for a Marxist at least—a considerable revision of at least the explanatory dimension of Marxism is on the agenda.

III

What does this slice of Marxist intellectual history mean to one concerned with the practical problems of the moral crisis of the late twentieth century? Let me first describe a context and then put my answer within it. As far back as the French Revolution at least there were serious thinkers who saw the need for a new basis for ethics. As E. J. Hobsbawm reminds us,

---

[49] Ibid., p. 113.

The post-revolutionary generation in France are full of attempts to create a bourgeois non-Christian morality equivalent to the Christian: by a Rousseauist "cult of the Supreme being" (Robespierre in 1794), by various pseudo-religions constructed on rationalist non-Christian foundations, but still maintaining the apparatus of rituals and cults (The Saint Simonians and Comte's "religion of humanity").[50]

And in Germany, I think that Hegel's *Phenomenology* can be fairly described as an attempt to develop a new morality in an atomistic and agnostic age.

None of those early pseudoreligions prospered. Nazism and Stalinism, which share much in common with them, did, for a while at least. But in the century and a half between the French Revolution and the 1930s, Marxism was the first mass movement committed to atheism that tried to face up to this issue in a democratic and thoughtful way. So, I would argue, the history I have just outlined, for all of the musty scent of a naive past that clings to it, poses issues we have not yet resolved. And insofar as it deepens the concept of a Marxist view of ethics, perhaps it might even help speed that resolution. But before turning to that possibility, a more recent—living—history should be noted.

The most fateful development for Marxism in the twentieth century was the triumph of Stalinism in Russia. And this had very serious consequences for the Marxist consideration of ethics.

Stalin took over the scientific, evolutionary, Darwinian version of Marxism first elaborated by the socialists of Kautsky's generation and put it to a profoundly antisocialist use. For the Kautskyans, whatever their limitations, Marxism was the integrating ideology of a democratic workers' movement in Europe. Its scientific claims, as Gramsci acknowledged, sometimes had an ideological "aroma" and gave a sense of "necessity" to struggles for daily bread, yet it was deeply an-

---

[50]*The Age of Revolution,* 1789–1848 (New York: Mentor Books, 1962), p. 260.

chored in people's lives.[51] Stalinism, on the contrary, liquidated all of the popular conquests of the October Revolution and initiated a top-down totalitarianism. One of its ideological justifications was that Stalin, as the supreme Marxist and therefore the supreme scientist, had the right to impose the good of the working class upon the working class. (It was not, of course, put in such crass and frank terms, but that is what it amounted to.) In ethics it led to the equation of morality with the most recent dictate of the omniscient leader.

This was the era not simply of Stalinism but of Nazism and World War II as well. The scholarly debates that were possible under conditions of relative capitalist stability prior to World War I were no longer possible. The democratic Marxists of the Second International, with some notable exceptions (e.g., the "Austro-Marxists," who continued to deepen their theories), either fought a rearguard action against the Stalinization of Marx or else began to retreat to a non-Marxisn socialism in which Marx was recognized as one precursor among many. Stalin died in 1953 and Khrushchev gave his famous speech on his predecessor's crimes in 1956. That obviously marked a great historical shift, and it was accompanied by action as well as theory—the Hungarian Revolution and the October events in Poland in 1956 and later by the Czechoslovakian spring of 1968. The Marxist ethical vision of freedom thus became a political force.

To be sure, something like a Marxist underground had survived even in the years prior to 1956. The Frankfurt school in Germany had been a center for genuine Marxism before and after World War II; in Yugoslavia, after the break with the Soviet Union, a philosophic school made up of veterans of the partisan struggle and committed to an authentic Marxism emerged. In France there were a number of intellectuals and groups carrying out a similar task. For example, in 1948, Maximilien Rubel published his *Pages from Karl Marx for a So-*

---

[51] Antonio Gramsci, *Quaderni del Carcere,* vol. I (Turin: Einaudi, 1966), pp 12–13.

cialist Ethic. For Rubel, Marx's "thesis on the inevitability of socialism is in the domain of those truths which, to become 'objective,' require the active *participation*, ethical *engagement* on the part of people (the second thesis on Feuerbach)."[52]

Indeed, Rubel pointed out that Marx himself had all but formulated just such a dialectical synthesis of the objective and the ethical. In a letter to Engels in 1867, he playfully reviewed *Das Kapital* as he imagined a Leftist editor, Karl Mayer, would have done it. Mayer, Marx projected, would approve of the "positive developments" of the book with regard to the division of labor, the machine system, and so on. But, Mayer would continue, the "*subjective* tendency" of the author was something else and his visions of socialism had nothing to do with his analysis.[53] It is clear that Marx was poking fun at this notion of a "value-free" analysis, a notion which, ironically, is so often attributed to Marx himself.

However, it was not only the work of men like Rubel but also the shock of the Khrushchev speech that paved the way for a renaissance of interest in ethics among Marxists. Within the various communist parties and on their margin, a reconsideration of Marxism, with a stress on its activist and ethical component, began to take place. The discussion then spread throughout various socialist parties, the Trotskyist and Maoist groups, and even among non-Marxist academics. But how does one evaluate this development? And how does it touch upon the theme of Marxism and ethics?

Leszek Kolakowski was once a leading revisionist in Poland, but since his effective exile from his native land he has increasingly moved away from any kind of Marxism. So his interpretation of these events is of more than mere scholarly interest.

The new Marxists, Kolakowski writes, attacked the "reflection" theory of cognition as developed by Lenin (our ideas merely "reflect" material reality) and stressed the interaction

---

[52] *Pages from Karl Marx for a Socialist Ethic*, 2d ed. (Paris: Payot, 1970), p. 30.
[53] Ibid., pp. 31–2; *MEW*, vol. 31, pp. 403–404.

of subject and object. In the same spirit they criticized the deterministic theories of communism that left out will and chance. Finally, he says,

they criticized attempts to deduce moral values from speculative historiographical schemata. Even if it were supposed, wrongly, that the socialist future was guaranteed by this or that historical necessity, it would not follow that it was our duty to support such necessities. What is necessary is not for that reason valuable; socialism still needs a moral foundation, over and above its being the result of "historical laws.[54]

Kolakowski, who was himself one of these new Marxists who helped to elaborate this view, now rejects it.

Attempts to combine Marxism with trends originating elsewhere [i.e., opening up Marxist scholarship to the rest of the modern world] soon deprived it of *its clear-cut doctrinal form;* it became merely one of several contributions to intellectual history, *instead of an all-embracing system of authoritative truths among which, if one looked hard enough, one could find the answer to almost everything.*[55]

But this is the destruction of a straw man. The "integralist" Marxism which Kolakowski describes was always a delusion, whether in the naive version of a Kautsky or the sinister and totalitarian version of a Stalin. Any thinking Marxist must welcome the disappearance of such an "all-embracing system of authoritative truths." But does that mean that Marxism *as a way of thinking* has lost a special relevance to the analysis of the ethical crisis and other problems of the late twentieth century? I think not (and will try to briefly summarize that relevance shortly).

A second attitude is that of sophisticated neo-Stalinism. That, for instance, is the point of view of Hans Jorg Sand-kuhler, the coeditor of the very useful collection *Marxismus*

---

[54] *Main Currents of Marxism,* vol. 3 (Oxford: Clarendon Press, 1978), p. 462.
[55] Ibid., p. 465. Emphasis added.

und Ethik. For Sandkuhler, the neo-Kantians, and the broader discussion of ethical questions in the pre-World War I socialist movement simply expressed the presence of a growing number of petty bourgeois people in the ranks of the party. This is, to begin with, one of those genetic arguments which pretends to have evaluated a development once it has explained its origins. Engels was a businessman, Marx an unemployed Ph.D., which proves . . . what? Indeed, Lenin himself insisted—along with Kautsky—on the importance of nonproletarians in bringing proletarian consciousness to the workers.

But much more importantly, the use of this terminology simply overlooks one of the most profound social class developments of the recent past. In the *Communist Manifesto,* Marx and Engels had written about the growing "simplification" of class relationships, with capitalism splitting into a tiny bourgeoisie and a huge working class with fewer strata. In their own lifetimes Marx and Engels realized they had been wrong and tried to correct their earlier view—only they did so in works that are much less widely known than the *Communist Manifesto.* But whatever their thoughts on the matter, events since Engels's death in 1895 have certainly demonstrated beyond the shadow of a doubt that the intermediate strata are of enormous importance. Serious Marxists, like Nicos Poulentzas, have tried to grapple with the resulting problems.

Under such social and economic circumstances—when the petty bourgeoisie is proliferating and the proletariat contracting (relatively, if not absolutely)—it is not intellectually serious to repeat stylized excommunications about the petty bourgeoisie and its pernicious idealistic tendencies. Clearly, the original Marxist model of the relationship between class and ideology has to be revised. And when Sandkuhler refuses to do this, but simply quotes various Soviet bureaucratic moralists as the truth counterposed to petty bourgeois Kantianism, that only further debases Marxism. When socialism was seen as the political—and cultural, ethical—task of a vast working-class majority, one could speculate on the emergence of a uniquely and specifically proletarian morality as the normative basis of the new society. But the new class structures clearly require profound revisions in the original vision.

A more serious attempt to preserve a coherent if not seamless Marxism was made by Perry Anderson, a British socialist who is a very thoughtful representative of Trotskyist Marxism. In his brief study *Considerations on Western Marxism*, Anderson attempts to summarize both the "underground" and post-Stalinist Marxisms.

For Anderson, "Western Marxism" includes figures like Lukacs, Korsch, Marcuse, Adorno, Goldman, and Colleti. Classical Marxism, Anderson argues, was central or eastern European, but these figures are mainly from the West.[56] Moreover, unlike the pre-World War I socialists or the Bolsheviks of the Revolution itself, these thinkers have worked within a context in which theory and practice are more and more separated from one another.[57] Because of this, the Western Marxists—many of whom were professional philosophers rather tha professional revolutionaries—are idealist, Hegelian, and oriented toward Marx's early writings. They are concerned with "superstructure" rather than "base" because they are heirs of defeat.

In his original manuscript, written in 1974, Anderson had written,

Meanwhile the change of temperature since the end of the sixties has also had its effects on Western Marxism. The eventual reunification of theory and practice in a mass revolutionary movement, free of bureaucratic trammels, would mean the end of this tradition.[58]

Clearly, Anderson at this point looked forward to precisely such a reunification. But in an afterword, written in 1976, he stated some reservations about his own revolutionary hopes.

I think Anderson's reservations are right (indeed, I would go beyond them). The complexity of society is now such that

---

[56] *Considerations on Western Marxism* (London: New Left Books, 1977), p. 26.
[57] Ibid., pp. 29, 92.
[58] Ibid., p. 101.

the unification of theory and practice as imagined by Marx and the early Marxists is not on the agenda. Rather, as I will detail in the next section, a long, difficult, and ambiguous time lies ahead *if* there is to be a transition to socialism. This means that the confident assertion of the privileged—analytical and ethical—position of the Marxist perspective, say, the one made by Lukacs in *History and Class Consciousness,* has to be abandoned. Bourgeois morality is certainly exhausted and all but contentless, but it is absurd to argue that a socialist ethical alternative is, or will be in the immediate future, clear. Is there, then, *anything* left of the Marxist perspective on ethics? I think so.

IV

In considering the positive contribution a Marxist approach to ethics might make in the current period, the work of Antonio Gramsci is of great importance. In significant respects, Gramsci deepened (and corrected) Marx. He stressed that the transition to socialism would be much longer than Marx had thought. In the giddy, euphoric time just before the revolution of 1848, Marx thought that socialism would come in a matter of years. But by 1850 he had changed his perspective radically and argued that it might take half a century before the great change would occur.

Gramsci realized that even the revised estimate was too optimistic. He changed the Marxist timetable, in part because he reconceptualized the socialist transition itself. That transition, he argued, did not only concern economics and politics but culture and society as well. During an entire historic epoch there would be battles for "hegemony" in the field of ideas and values. The revolutionary task was to transform prevailing modes of thought and feeling, and that could not be accomplished in a few years or even in a single generation.

Gramsci died in the 1930s, a victim of Fascism. Yet even though he developed a version of Marxism radically different from that of his contemporaries, Kautsky and Lenin (he opposed the "scientistic" Marxism of both of those thinkers), he still believed in an integral system.[59] If there is to be a transition to a truly humane socialism—and not, as is quite possible, the emergence of authoritarian collectivism on a world scale—we must nourish the spiritual and cultural emancipation of which Gramsci spoke while being even more open, more sensitive, and less dogmatic than he was.

Marx and Gramsci shared, I think, many important, fundamental values—above all the commitment to the self-emancipation of the working class and, through it, the universal emancipation of mankind as a whole. Gramsci made Marxism more profound in the light of experiences that came after Marx's death (Gramsci was born eight years after Marx died), but he did so in a profoundly Marxist way. And this relates to our present plight.

The characteristic of the late twentieth century that most touches upon any consideration of morality is the ubiquitous spread of relativism. I will turn to Hegel—a "Marxist" Hegel, if you will—to help explicate this point.

In his unforgettable account of the Enlightenment in the *Phenomenology,* Hegel paints an extraordinarily dramatic picture of the cultural development that took place during the eighteenth century. The Enlightenment's intellectualism, he said,

spread like a scent in an atmosphere which offered no resistance to it. . . . [an] invisible spirit which no one remarked spread through all the high-minded parts [of faith] and took over the very bowels and members of the unconscious gods and *"on one fine morning* gave the comrades a shove with the elbows and, Bautz! Baradautz! the gods were lying on the ground.[60]

This process has proceeded apace. Like a scent, relativism has spread through Western society and practically no one knows what they believe any more.

[59] Cf. Leszek Kolakowski, *Main Currents of Marxism,* vol. 3, chap. 6.
[60] *Phenomenologie des Geistes* (Frankfurt: Suhrkamp Verlag, 1975), pp. 402–403. Here Hegel quotes from Diderot's *Rameau's Nephew.*

This result is analogous to the situation that Hegel, in an earlier work, had described among the Jews on the eve of the emergence of Christ:

The spirit is gone out of the constitution and vanished from the laws and because of this change no one believes in the laws. So there comes a seeking, a striving, for something else, which each finds in something else and thus there is a tremendous variety of cultures [*Bildungen*] of life styles, of claims and needs which, if they cannot survive in peace alongside of one another, finally break up the old system and give a new universal form, a new interconnection to people.[61]

I do not know if Hegel was right about Jewry in that period; but his words strike me as perfectly appropriate for the late twentieth century.

The problem is, where will the basis of the "new universal form" be found? Not, I believe, in any kind of authority or revealed truth. This is not to say that revealed truth is false but only that it is no longer capable of playing the social consensual role it once did. If there is to be a new morality, it will have to be shaped within an agonistic civilization—no other kind of civilization is possible in this age. It will be a skeptical morality, but it will also have to recognize some enduring, nonrelative values. We need, as an Italian Marxist Antonio Banfi has said, a

Copernican man . . . for whom there is no metaphysical outlook, who acts as a part of nature, who completes his own historical work and so achieves the universality of consciousness and the rationality of knowledge which is facilitated through his practical, historical effects.[62]

This is not to argue for a new messianism but for a radically transformed consciousness of "ordinary" people. Some of the most poetic and soaring images of the Marxist aspira-

[61] *Der Geist des Christentums,* in *Werke,* vol. 1 (Frankfurt: Suhrkamp Verlag, 1971), p. 297.
[62] Quoted in *Geschichte des Marxismus,* vol. 2 (Frankfurt: Suhrkamp, Pedvag Vranicki, 1972), p. 952.

tion—the average citizen as Renaissance man, as in the young Marx; Trotsky's anticipation of numbers of geniuses who would rise above Aristotle and Shakespeare—have not only turned out to be romantic and wrong but capable of rationalizing totalitarian "new men." What is needed now is a mass ethic for a society of deracinated, bewildered pleasure worshipers. That is a more audacious, if more prosaic, project than the one imagined by the young Marx and by Trotsky.

What does Marxism have to do with this need? Its relevance, I think, lies on the plane of high theory: in the fact that it is both relativist and absolutist, that it analyzes and proposes the interpenetration of those seeming opposites. And, on a more mundane level, Marxism, if properly understood, focuses on some critical interconnections between morality and society.

First, there is the dialectic of the relative and the absolute. In 1923 the young Lukacs remarked that for Marxism

the absolute is not abstractly denied but rather conceptualized in its concrete, historical form [*Gestalt*] . . . because the historic process in its uniqueness, in its dialectical forward movement and in its dialectical backsliding, is an unbroken struggle for higher levels of truth which (socially) are the self-knowledge of humans.[63]

More recently, Karel Kosik expressed a similar view in *The Dialectic of the Concrete,* one of the most powerful contemporary Marxist analyses:

For the a-historical view, the absolute is non-historical and eternal in the metaphysical sense; for historicism, the absolute does not exist in history; but for the dialectic, history is a unity of the absolute in the relative and the relative in the absolute, a process *in which the human, universal and absolute emerges in the form of a general precondition as well as in the form of a specific historical result.*[64]

But then what is this relative absolute to which these Marxists refer? Man, of course. The specifically human—hu-

[63] *Geschichte und Klassebewusstsein,* p. 206.
[64] *Dialektik des Konkreten* (Frankfurt: Suhrkamp Verlag, 1976), pp. 142–43.

man nature in the sense described in the first section—develops through history, through struggle, through choice. The conquest of more space for that human nature is the purpose that can bind a society together. In religion, Durkheim once remarked, man always worshiped himself. But now religious man as a social, political actor is finished and there is only man.

In the United States today there is a pervasive individualistic hedonism—which is related to the ubiquitous relativism—among the affluent and the not so affluent. As Adler so prophetically intuited, there could be—and there is—a society that is less materially oppressive than the cruel orders of the past, but it smothers genuine freedom. As a corollary to this, Adler added that one of the critical elements in changing such a situation is precisely the existence of a socialist ideal as ideal. Socialism is a necessary but not a sufficient condition for the assertion of human freedom—which is to say, of a morality under conditions of mass atheism.

Keynes once gave a toast to the Royal Academy of Economics. "To the economists," he said, "who are not the trustees of civilization but the trustees of the possibility of civilization." To the Marxists, I would say the same.

CHAPTER FIVE

Marxism and Ethics Today

ELIZABETH RAPAPORT

INTRODUCTION

Michael Harrington has rescued Marxism from some of the most entrenched distortions perpetrated by Marx's critics and by Marxists from whom those critics learned the Marxism they reject. Harrington and I are in agreement on a number of broad but crucial questions about the nature of Marxism and of a Marxist approach to ethics today. My Marx too is a moralist, a democratic humanist, and a subtle social analyst rather than a doctrinaire economic determinist. I will follow Harrington's lead in identifying myself and my Marxism politically in order to clarify our different interpretations of Marx.

The tendency within the socialist movement that most authentically represents Marx's socialism is that of the anti-Stalinist revolutionary left. This socialism differs from Harrington's in fatefully and fully linking socialist values and transformation to the development of a revolutionary class, the proletariat, and in priding itself on the power of Marxism as a scientific organon. This Marxism has misunderstood the development of the proletariat and made exaggerated and erroneous claims for the power of Marxian science. Nonetheless, it is the Marxism of Marx and no assessment of the value of Marx's work today can be made unless we acknowledge the flaws as well as the strengths of his work, as Harrington has always insisted in his writings.

Having heard this much, Harrington will not be surprised to learn that I am that irritating brand of socialist, the revolutionary pessimist or skeptic. Neither of us regards socialism as either inevitable or impossible. But I regard the extent to which history has proved Marx wrong to be a much deeper blow to the capacity of socialists to understand and implement social change than does Harrington. I am trying to listen to social democrats such as himself with more openness than heretofore. But when I listen I still hear Roberto Michels's maxim: "Reformism is the socialism of nonsocialists with a socialist past."[1] I cannot envision social democracy as the road to the classical Marxian socialism that Harrington and I understand in much the same way. I see social democracy as a defensive strategy, as the only available counterweight to a rightward shift in the no-growth or declining America whose measure we are just beginning to take.

Harrington will again not be surprised to learn that my main quarrel with his interpretation of Marx is that he is turning Marx into Feuerbach, he is reverting to utopian or ethical socialism without apparently appreciating the questions this raises about the worth of the whole Marxian corpus.[2] "Utopian socialist" was the term Marx and Engels employed to describe those who postulated socialism as an ethical ideal.[3] They dis-

[1] In 1911, Roberto Michels published a path-breaking, extremely influential study of the oligarchic tendencies inherent in socialist as well as any other organization. The quotation is from this work, *Political Parties*, ed. Seymour Martin Lipset (New York: Free Press, 1966).

[2] See Harrington, p. 86. Ludwig Feuerbach's critique of Christian theology in the late 1830s and early 1840s is amongst the most important influences on Marx's youthful socialist views. In 1843 Marx proclaimed himself a Feuerbachian and a socialist. Feuerbach argued that human beings should make a Christian community on earth rather than seeking a heavenly antidote to life's disappointments. He identified this goal with socialism. He did not develop any political program for the achievement of socialism or interest himself in how this earthly Christian community was to be achieved in practice. A collection of Feuerbach's most important writings is available in Z. Hanfi's edition, *The Fiery Brook* (New York: Anchor/Doubleday, 1972).

[3] See the jointly authored *Communist Manifesto* and Engels's essay "Socialism: Utopian and Scientific," a widely available excerpt from his book *Anti-Dühring*.

tinguished themselves from the utopian socialists on the grounds that while they too subscribed to socialism as an ethical ideal, they also (1) advocated proletarian revolution as the means of bringing about socialism and (2) had developed a scientific understanding that permitted them to predict socialist revolution. Harrington differs from those whom Marx and Engels called utopian socialists in that he is deeply committed to the pursuit of socialist aims by political means, while the classical utopians expected capitalism to collapse before the sheer moral power of the socialist critique. But he diverges from Marx and reintroduces utopianism in three respects: (1) He abjures any claim that Marxism has the scientific capacity to predict the emergence or the contours of socialism. (2) Rather than base socialism's political future on a proletarian movement, he looks to an alliance of disparate classes. (3) He envisions the introduction of socialism through a series of gradual reforms rather than through revolutionary confrontation with capitalism. I cannot resist asking him this question: If the mature Marx was a social democrat, why did not the Second International know it? The Second International was founded in 1889 when Engels was still alive and endured until World War I. Why did they cling to revolutionary rhetoric even while their practice was reformist if they could have cited Marx and Engels as fellow reformists? That the Stalinist Third International would obscure their differences with Marx in this respect is perfectly plausible. Why did the Second International obscure their affinities?

In responding to Harrington I will try to do three things. I will sketch what I take to be the chief claim of Marxism on the attention of ethical theorists today, the areas in which it clearly has something to teach. Second, I will delineate what I take to be its greatest present internal weakness as a philosophical program. And, third, I will provide an example of a substantive Marxist analysis of contemporary society and suggest that while there is much revisionist work yet to be done, Marxism remains a vital and fruitful mode of critique.

I

In his paper Harrington very accurately diagnosed the current ethical and spiritual crisis. "Everyone . . . has become, existential, situational and the like. But that provides no basis for solidarity, for a common, consensual normative consciousness."[4] Yet contemporary moral philosophers often ignore this very problem of a lack of common standards when analyzing the social policy questions that this lack renders so intractable. In the past ten years American moral philosophers have executed an astonishing *volte-face* on the question of whether philosophers have a role as advocates in public policy debates. The standard answer to this question a decade ago was that philosophers were pecularily qualified to analyze the logic and meaning of moral discourse but were in no way privileged in their ability to make correct moral judgments. This doctrine was a straightforward application of the then equally standard (but of course not universal) trichotomous fact/value/analysis distinction. Moral discourse was divided from scientific discourse and philosophy from both. Today philosophers are more than willing to take a stand on public issues, abortion, euthanasia, violence as instrument of social change, any element of U.S. foreign policy, preferential treatment of previously discriminated against social groups, and so on. This reversal is easy enough to account for historically. The proximate causes are well known: the return of public debate and the patent lack of consensus on both domestic and international issues that emerged in the middle sixties; the blank refusal of a new generation of students to believe that this aloofness was indeed a professional virtue in their teachers; and the commitment and activism that the new era called forth in teachers as well as their students. These, of course, were not changes in our conception of what philosophy was or how it was properly done. The practice of advocacy requires that we revise our conception of philosophical method. Both the valuable results of the

---

[4] See Harrington, p. 89, this volume.

current advocacy literature and its limitations compel this conclusion.

Let me first describe what I take to be the typical procedure of the philosopher as advocate. The philosopher usually starts out with a commitment (e.g., to the preposition that a pregnant woman has an unrestricted right to an abortion). The philosopher attempts to support this judgment with cogent and conclusive arguments. Typically, there will be no attempt to establish the first principles of morals, to argue for or against utilitarianism or any other full-fledged normative theory. What we look for are premises that are rich enough to yield the desired conclusion and that are consensual—not themselves subject to controversy. The philosopher will have another string to his or her bow. An assault will be made on other philosophers who have exhibited faulty logic or conceptual analysis in their attempts to argue that well-entrenched norms sustain alternative value judgments and policy choices. I believe that we can summarize the results of these efforts as follows:

1. These investigations have revealed that, in many of the most important public debates being conducted today, there are no consensual norms to which we may appeal. The value of much of the philosophical work that has been done lies not in its success with its explicit program of providing morally sound solutions to public policy debates but in revealing that there is no consensus and in making explicit what the crucially contested norms and concepts are.[5]
2. If the foregoing claim is correct, it follows that philosophers must either give up advocacy or adherence to the conception of philosophical method which denies philosophy can adjudicate substantive normative disagreements. Our practice to date has for the most part not faced up to this choice.

[5] See Alasdair MacIntyre, "How Virtues Become Vices," in *Evaluation and Explanation in the Biomedical Sciences,* ed. H. T. Engelhardt and Stuart Spicker (Dordrecht: Reidel, 1974); and Elizabeth Rapaport and Paul T. Segal, "Abortion and Ethical Theory," in *Sex: from the Philosophical Point of View,* ed. R. Baker *et al.* (Totowa, N.J.: Littlefield Adams, 1977).

We have sought to operate in the formerly forbidden domain
of moral judgment proper with nothing but the tools of
conceptual analysis, clarity, and rigor. But as every adher-
ent of the traditional fact/value/philosophy distinction knows,
substantive disagreements cannot be settled by these means.
3. If we are to engage in advocacy, what is needed is a critical
ethics. A critical method would facilitate the location of
important normative divisions in our social world as well
as help us to understand how they might be overcome.

If we examine the recent and the not so recent history of
Anglo–American moral philosophy, I think we will find that
there is a broad consensus on the nature of method in ethical
theory. Its deepest assumption is that there is a moral consensus underlying the apparent disagreement in moral life. Its
most characteristic conception of the purpose of ethical theory
is to reveal the structure of ordinary morality by eliciting its
implicit and unifying principles. Therefore the most characteristic methodological test of normative theory is agreement
with our presystematic moral convictions. This methodological
stricture has been adopted by the school of moral philosophy
most apparently revisionist in its intentions, the utilitarians.
They too accept that utilitarianism is to be tested by its fidelity
to our clear and distinct moral ideas.[6] Some formalists admit
there are limits to moral consensus.[7] But this is treated as a
brute, irremediable fact about moral life, not an occasion for
revision.

In accordance with this methodological perspective, there
are two proper occasions for the revision of presystematic ethics, two interventionist employments of ethical theory. With
the help of an ethical theory that reveals the animating struc-

---

[6] An exception is J. J. C. Smart. See "Extreme and Restricted Utilitarianism,"
Philosophical Quarterly 6 (1956). He does not elaborate an alternative position but he does repudiate fidelity to common conviction as a criterion of
theoretical adequacy.

[7] Notably, R. M. Hare. See *Freedom and Reason* (New York: Oxford University
Press, 1963) on the problem of fanaticism.

ture of ordinary morality, greater consistency and completeness may be induced on ordinary morality. Where inconsistencies obtain, those judgments that are inconsistent with the principles revealed by theory may be eliminated. Where ordinary morality finds some moral questions puzzling and is unsure what course is correct, the possession of a clear comprehension of its underlying principles may make it possible to fill out ordinary morality, to extend the area of moral clarity. It is apparent that if endemic conflict is a central feature of moral life, this methodology is inadequate. If conflict rather than consensus in moral life is taken as the guiding assumption, then the question of what controls on theoretical revision are to replace the sanctity of putatively common conviction becomes the central methodological question of theoretical ethics.

If philosophers assume disagreement rather than consensus, what mode of investigation available to them would warrant the recommendation of some value commitments and social policies rather than others? Marxism is a mode of analysis that recognizes and responds to the existence of value conflicts within society. Unfortunately I can do no more here than refer to historical materialism as a framework for the analysis of value conflict and value change.[8] Harrington is so concerned to defend Marxism from vulgar mechaninist readings that he all but denies that there is to be found in Marx a theory of history, properly called historical materialism, which is in part a method—not a recipe—for providing an account of value conflict and change.[9] This method has been ill understood by Marxists for a myriad of reasons ranging from the sketchy nature of its presentation in Marx's writings to the positivistic climate of thought in the generations succeeding his to the sheer ineptitude for which some of his epigones are

[8] See my essay "Ethics and Social Policy," *Canadian Journal of Philosophy* 11, no. 2 (June 1981): 285. For a brief and accurate account of historical materialism see Anthony Giddens, *Capitalism and Modern Social Theory* (New York: Cambridge University Press, 1971).
[9] See Harrington's "The New Karl Marx," in *The Twilight of Capitalism* (New York: Simon & Schuster, 1976), pt. I.

justly scorned. Further, historical materialism almost certainly does not permit fully adequate answers even to the questions it raises about values—to say nothing of value questions outside its scope. It presses the assuredly large explanatory and shaping role of class and class conflict further than is sustainable. To name only the most obvious counterinstance, a purely Marxist account of relations between the sexes is a heroically misconceived project. Nonetheless, Marxism directs our attention to value conflict, and—as will be shown by example in section III of this essay—can account for aspects of these important phenomena whose very existence is obscured by the habits of thought characteristic of analytic philosophy. In the next section I will address what I take to be the central problem plaguing those who would make use of a Marxist methodology today.

II

Marx made three related critiques of his contemporary world: A critique of bourgeois society, a projection of socialist values to answer to the defects of bourgeois society, and the more distant projection of classless communist society. I will refer to the whole complex critique and projection as Marx's socialism. Like Harrington, I am deeply responsive to the core values of classical Marxian socialism, the Promethean values of freedom and solidarity realized through the creative transformation of self and nature. However, I must insist that Marxists, if they are to contribute to the current debate about ethics, have a much larger task to perform than Harrington acknowledges. Harrington is content to cleanse Marx of the misinterpretations of Stalinists and other mechanists, of the dead weight of false predictions, and what he takes to be an undeserved reputation for trading in historical inevitabilities. Marxists must confront a deep crisis in Marxist methodology that has been latent since the Second International. Harrington is well aware of this crisis. He sees it largely as an affliction to which scientistic misinterpretations of Marx have exposed us, while I see it as an epistemological crisis within authentic Marxism.

In his book *Socialism,* Harrington identifies the most serious challenge to the relevance of the classical Marxian vision of socialism today. It is Marx's single greatest mistaken prediction: That "the working class was compelled by the conditions of its existence to struggle for socialism."[10] While he recognizes that history has forced us to repudiate Marx's scenario of working-class development, Harrington does not appreciate that to repudiate this scenario leaves him in the position of advocating a purely ethical or utopian socialism. What is wrong with a purely ethical socialism? Let us look at this question from Marx's own point of view. It renders the Promethean values of classical Marxism just one more among the plethora of what Harrington correctly describes as the "existential"[11] choices available in our contemporary secular, pluralist world. They are sanctioned by neither the historical allegiance of the working class, science, nor any other plausible warrant-providing ground. Nor does it preserve the classical Marxist connection between what is possible and historically emergent social patterns. Like Harrington, I think this connection has been exaggerated by Marxists. But Marx was the founder of this Marxist exaggeration. Nothing was closer to Marx's ambition than to establish that socialist values were fully grounded in emerging historical trends.

A Brief Sketch of Marx's Career as a Socialist Critic

Marx's career as a socialist critic may be divided into two phases, his early or Feuerbachian phase and his second or historical materialist phase. In his Feuerbachian years of 1843 and 1844, he marshalled and expanded the resources of Feuerbach in order to demonstrate that socialism was the answer to three salient deficiencies of capitalist society: (1) democracy would only be a sham without the abolition of private property; (2) labor and all other human activities were "alienated" in capitalist society; and (3) the proletariat, even when granted full democratic rights within capitalist society, was so system-

---

[10] *Socialism* (New York: Saturday Review Press, 1970), p. 351.
[11] Harrington, p. 89, this volume.

atically reduced to a poor and bestial existence and so deprived of the power to improve its position that to emancipate itself it would have to destroy capitalism. Thus, it was the engine, the "heart" of the socialist revolution.[12] *Mutatis mutandis*, these three elements remained the hallmark elements in Marx's normative theory of socialism. But in 1845 Marx repudiated the Feuerbachian methodology upon which he had relied in formulating this early critique.

The repudiation of Feuerbach created an epistemological crisis for Marx, which he resolved by developing historical materialism. It will be useful in tracing out the development and present status of scientific Marxism if I adopt the overdrawn but not misleading interpretative thesis about Marx's career that he developed historical materialism *in order to* (although not *only* in order to) defend and extend through better science his now unsupported pre-1845 socialist critique. Marx's solution to this 1845 epistemological crisis set the stage for a second crisis in Marxist epistemology which has been latent since the days of the Second International. In essence, Marx repudiated Feuerbachian essentialism. Marx had taken over Feuerbach's concept of the *Gattungswesen*, the species being, the truly or fully human nature which supplied critical normative standards against which the deficiencies of capitalist life could be measured and in accordance with which appropriate alternative institutions could be envisioned.[13] In 1845 Marx was assailed by the historicizing insight that there was no essential human nature to provide a critical standard for judging all social epochs. Human "essences" are historically local and historically changing, the product of patterns of human practical activity. Let us look at historical materialism as an attempt to solve the following dilemma of post-es-

[12] See Marx's writings of this period, "On the Jewish Question" and "Towards a Critique of Hegel's Philosophy of Right," available in *Karl Marx: Selected Writings,* ed. David McLellan (New York: Oxford University Press, 1977), as well as in other accessible collections.

[13] See in particular Feuerbach's *Essence of Christianity*, relevant selections from which are to be found in the Hanfi edition cited above.

sentialist normative theory: If there is no theory of essential human nature to provide a critical standard according to which we can hope to assess and transform current social practices, where may we look for such a critical standard? Where the Feuerbachian Marx appealed to the concept of the *Gattung-swesen*, the historical materialist discovers the content of socialist norms in the emergent forms of the socialist mode of production and the proletarian consciousness which is shaped by them and by the political struggle to overthrow capitalism. Taking the historical materialist overview in conjunction with Marx's theory of capitalist development, it is possible to construct a normative theory of socialism as a projection of the consciousness of the revolutionary proletariat, the emergence of which that theory predicts. The content and amplitude of that normative theory will depend upon and be limited by the predictive capacity of socialist science.

At this point it may well look as though I were attributing to Marx a spontaneist view of class consciousness, implying that when the time for revolution is ripe, the working class achieves a sufficient grasp of its objectives and its appropriate strategy without the mediation of theory or political organization and leadership. Marx did not hold a spontaneist view of class consciousness. But it is very important to distinguish between Marx's nonspontaneist theory of the formation of class consciousness and the kinds of nonspontaneist positions that Marxist apologetics have concocted if we are to understand why the classical historical-materialist solution to the problem of explicating socialist norms fails.

Class Consciousness and the Latent Crisis in Marx's Epistemology

Marxists typically refer to three quite different things when they speak or write about proletarian class consciousness: (1) the values and beliefs, particularly the political values and beliefs, of the working class or sections of it; (2) a working-class party, its program and organization, the militant embodiment of class consciousness; and (3) socialist theory (e.g.,

Marx's critique of economics or theory of the transitional socialist state). It therefore is possible to dispute not only *what*
"true" class consciousness is doctrinally but also *who* possesses
it. Answers to this question have included the workers, the
militant workers, the vanguard party, the inspired leader, the
living Marxist theoretician, and the dead Marx. Marx's own
answer to this question as I reconstruct it combined and required all three elements. While he certainly did not believe
that every worker would master *Das Kapital* or that political
organization was supervenient, he believed that revolutionary
socialist ideas and affiliations would, in time, be as irresistible
to the proletariat as the attempt to breathe by one who is
suffocating. Had capitalist development taken the course that
Marx predicted, had what I will call the "classical scenario"
eventuated, it is entirely plausible to suppose that such a revolutionary consciousness would have emerged. In my view, if
anything should be understood to be inevitable within classical
Marxism, it is the emergence of a revolutionary proletariat
under the conditions of capitalist development Marx envis-
aged. *Das Kapital* is the *bildungsroman* of the revolutionary
proletariat. Suppose generations go by and there is no such
emergent revolutionary proletariat?

Marx's theoretical successors were first made to face this
question in the Bernstein or revisionist debate that raged among
German socialists in the 1890s.[14] Edward Bernstein argued
that the development of capitalism had shown that the workers' position had been and could be further ameliorated—that
the Social Democratic party ought to adopt a strategy of class
collaboration and gradualism to replace the refuted theory of
class conflict and unnecessary proletarian revolution. The conclusion to be drawn from Bernstein's analysis was that proletarian consciousness was revisionist because the social world
of the proletariat was best captured by a revisionist analysis.
Karl Kautsky, to whose lot it fell, in virtue of his preeminence

---

[14] See Peter Gay, *The Dilemma of Democratic Socialism* (New York: Collier,
1952).

among German Marxists, to carry the argument against Bernstein, made three counterarguments: (1) Bernstein's views
were not orthodox Marxism; (2) Bernstein was wrong in many
particulars in his analysis of the non-Marxist course of capitalist development in the 1880s and 1890s and spectacularly
wrong, proving himself to be a moron, in failing to appreciate
that Marxist science revealed that the deeper tendencies of
capitalism were leading to chronic depression; (3) that working-class consciousness would only become sound revolutionary consciousness through the intervention of a revolutionary
party armed with scientific socialist theory.[15] The first coun-
terargument is, of course, contemptible. To castigate Bernstein
for want of orthodoxy ill becomes the scientific Marxist. Versions of Kautsky's second and third arguments have been the
mainstay of orthodox Marxism ever since. They had considerable plausibility in 1900 but have lost plausibility in every
succeeding generation.

The cogency of Kautsky's third argument requires the
cogency of his second argument. If it is Marxist science that
permits the revolutionary intellectuals to bring the working
class from its reformist or trade-union consciousness to revolutionary class consciousness, then that intelligentia has no
superior scientific understanding to offer unless Marxist science is able to correctly predict the decline or collapse of capitalism. Marxists who have refused to recognize the episte-
mologic consequences of the divergence between the classical
scenario of capitalist development and the historical reality
followed one of three courses. They have, beginning with Georg
Lukacs,[16] lost the fundamental insistence on concert between
theory and experience, the only thing that Marx and Bernstein
had in common, and confounded theoretical adequacy with
dialectical elegance. You can always tell that someone lacks

---

[15] It was from Kautsky that Lenin adopted the famous doctrine of *What Is to Be Done;* that is, the socialist intelligentsia must bring revolutionary consciousness "into" the proletariat from "outside."

[16] See George Lukacs's *History and Class Consciousness* (Cambridge, Mass.: MIT Press, 1971).

dialectical subtlety by the crass insistence that no theory, however esthetically satisfying, can be accepted unless it has some empirical resonance. The second course is to insist that Marx's economic predictions about capitalism will yet be vindicated and that, therefore, classical Marxist theory is still the best and only necessary guide for the socialist movement. While it is certainly possible and perhaps even likely that capitalism is now in the long-predicted irrevocable decline, it is a travesty for Marxists to suppose that a classical conclusion to capitalist history might take place without any impact being made by the actual history of capitalism and the proletariat that was shaped within it and helped shape it, that lessons need not be learned from a hundred years of reformism and Stalinism, that the overdue collapse of capitalism coupled with the ministrations of dedicated Marxists will willy-nilly transform the Western working class into revolutionary socialists. The third course is that taken by Harrington. Harrington is understandably so anxious to retrieve the good name of Marxism from the mechanistic determinists, the Stalinists, and other elitists that he makes a virtue out of what for Marx is a defeat: the failure of Marxian science to comprehend the lineaments of social development and the retrenchment to ethical socialism. I do not wish to exagerate my disagreements with Harrington. I do not wish any more than he does to revive scientific socialism. However, it seems to me to be an essential part of the rehabilitation of Marxism to come to terms with Marx's scientific aspirations and to mourn the failure of science to be able to provide the degree of insight and prediction he hoped to achieve, as well as to defend socialism and humanity against the pretensions and abuses of pseudoscientific Stalinist Marxism. Once this is done, we can begin to ask what the resources and potential of Marxian social criticism might be today to either defend classical socialist values or to engage in innovative socialist criticism.

III

I will close with a brief illustration of Marxist critique at work in a domain where its very potent resources are adversely affected by the unresolved crisis of its epistemic foundations just sketched. Marxist critique may be seen as relying on two component elements: the problematic Promethean values of emerging socialism and the critique of bourgeois values that appeals to no socialist values but rather exposes the internal incoherencies and deficiencies of bourgeois morality. It is this second aspect of critique that I wish to call attention to now. There is a pattern of argument which we encounter over and over again in Marx's writings whenever he has occasion to analyze "bourgeois right" (e.g., claims that the capitalist society can grant equal political rights to all citizens or that market capitalism compensates labor commensurately with its contribution).[17] Marx argues that capitalist society cannot vindicate in practice the values it claims to instantiate in its economic and political institutions. He further argues that, in time, socialism can both vindicate bourgeois right and supersede it with the superior Promethean values. These further constructive aspects of this pattern of argument I will ignore here. The full argument in the case of democracy would require showing not only that socialized property was a necessary condition for genuine democracy but that some available set of socialist political institutions were sufficient to realize genuine democracy. The full argument in the case of income distribution would have to show that there was at hand a nonexploitative form of socialism. These are obviously complex and elaborate projects. I will focus exclusively on the negative or destructive aspect of these antibourgeois arguments, the leg of the argument which purports to establish the claim that bourgeois society cannot vindicate its own values in practice.

Let me take as an example what I believe to be the holiest of holy principles of contemporary welfare-state capitalism. It is the meritocratic principle of distribution of positions within the social division of labor (i.e., that positions and educational opportunities with occupational import should be allotted to the best-qualified applicant, the identity of the best-qualified

---

[17] For examples, see *On the Jewish Question* written in 1843 and *Critique of the Gotha Programme* written in 1875. Both are in the McLellan collection cited above.

applicant to be established through fair and open competition).
Meritocrats will disagree as to what is required to make competition fair and open. Even the weakest interpretation of meritocratic justice defensible in current conditions is fiercely re-
sisted by millions of Americans who subscribe to older
conceptions of meritocracy. A full-dress Marxist analysis would
attempt to account for the historical radicalization of the meritocratic conception. It would map the various strata of our
population and the reasons why they tend to adopt certain of
the various weaker or more radical meritocratic positions. While
I cannot stay to defend it adequately here, the weakest possible
defensible meritocratic position under the conditions obtaining
in American society today is the one holding that fairness is
assured when sufficient social resources are expended on education to neutralize the effects of the unequal starting positions in life that result from differences in wealth and family
circumstances. This weak version of meritocracy will have to
give way to a more radical one unless the contingent claim
that such a remedial educational policy can achieve these results is borne out in practice.[18] More radical positions would
envision some system of preferential placement. I will present
a Marxist argument to the effect that capitalism cannot be a
genuinely meritocratic society.

Indeed, it is integral to this argument that even the clear-
eyed, honest contemplation of the depth of reform necessary
to honor meritocratic principles in practice is virtually impossible for bourgeois moralists, much less reform itself. Such
contemplation would necessarily undermine the most fundamental presuppositions of bourgeois morality. Avoiding this
consequence of critical thought explains the otherwise inexplicable lack of appreciation of the conclusion to which inquiry
into this set of issues logically leads. Here is the source of the
"profound hypocrisy" that Marx believed to be characteristic

---

[18] For the development of the progression of the distinct meritocratic positions
and what might force a meritocrat to move from a weaker to a stronger
version, see Thomas Nagel, "Equal Treatment and Compensatory Discrimination," *Philosophy and Public Affairs* no. 2 (1973).

of bourgeois thought.[19] The hypocrisy is not willful or cynical; it is the mechanism that protects vulnerable presuppositions which could not stand full critical scrutiny.

What bourgeois morality cannot contemplate but is nevertheless necessary to achieve a genuinely meritocratic society is the removal of *class* as well as race and sex as disabilities preventing people from achieving the places to which their merits under favorable conditions might be shown to entitle them. Were we to try to offer redress to so large a number—roughly three-quarters of our population is either working-class, black, or female—we would undermine the very premise necessary to defend meritocracy: that there is inevitably a shortage of educable talent to fill the relatively cherished places within the hierarchical division of labor. The presumptive problem of placement within a genuinely meritocratic system would be how to allocate places where there is a chronic oversupply of educable and qualified people. This is a problem that (classical Marxist) society could, in principle, not only face but would welcome because of such a society's commitment to reducing the hierarchical configuration of the division of labor over time.[20] But capitalist society cannot face this problem because it calls into question the deepest moral and sociological presuppositions of that society: that the hierarchical stratification of labor is natural or inevitable in an advanced industrial society. We cannot vindicate our claim to be a meritocratic society without neutralizing the normally decisive impact of class origin on life chances. We cannot do the latter with the kinds of reform strategies that are now or are likely to be taken as serious political options. If we did neutralize class origin as an unjust mitigator of opportunity, we would have to be prepared to contemplate an even more radical redesign of social institutions. I will not even raise here questions about whether any among these more egalitarian options might be practically feasible and morally desirable alternatives to a

[19] Harrington quotes Marx's phrase in connection with his discussion of British imperialism, p. 83.
[20] See Marx's *The Gotha Programme*, cited above.

genuinely meritocratic system. My sole claim is that, in the welfare state, the meritocratic gospel is utopian or sham—in either case, an illusion. The reforms necessary to introduce genuine meritocracy are not seriously contemplated. Nor does the shallowness of contemplated remedial policies or the degree to which we fail to be in practice a meritocratic society lead bourgeois moralists to draw the inescapable conclusion that we do not practice and show no signs of moving toward practicing what we preach. Whatever else Marxism can or cannot do in ethics today, it asks questions that allow us to see at least some of the limits and illusions of bourgeois morality.[21]

---

[21] An expanded version of the first part of this essay may be found in "Ethics and Social Policy," *Canadian Journal of Philosophy*, Vol. XI (2), June, 1981, 285–308.

CHAPTER SIX

Marx and Morality

ALLEN W. WOOD

Karl Marx wrote a great deal in support of his conviction that capitalism is an irrational, inhuman, and obsolete social system that should be overthrown. His entire social theory and practical endeavors, in fact, are focused on this conviction: on supporting it theoretically and acting on it effectively. Yet it is a striking fact that Marx said very little about the values in terms of which he denounced capitalism. Perhaps Marx exhibited an acceptance of certain values in the course of his attacks on bourgeois society, but he almost never said anything about what these values were or how they might be justified philosophically. The task of expounding Marx's "ethical views" is a treacherous one, partly because Marx had so little to say on the subject but also partly because he said too much. The little he did say suffices to refute most common interpretations of the "ethical foundations" of Marxism. While some of Marx's statements indicate his acceptance of recognizable, even conventional ideas, others clearly show that he held some novel, interesting, and extremely unconventional views about the nature of moral values and their place in social criticism.

It is easy to go far wrong in expounding Marx's ideas in this area; Professor Harrington has done an exceptionally good job of avoiding the traps into which most expositors fall. I do not agree entirely with him, but I am uncertain how far our disagreements are substantive and how far they are merely verbal. Let me begin by underscoring four important points that I think he gets largely right.

First, Harrington is right in denying that Marx agreed with the early Lukacs that the proletarian *party* is the repository of (that is, the infallible authority on) working-class consciousness, on what it is or what it should be. Marx *did* hold that the proletariat's class interest is not necessarily identical with what individual proletarians, or even the proletariat as a whole, at any given moment regards as its interest. Rather, Marx says, "it is a question of what the proletariat *is* and what, according to its *being*, it is historically compelled to do."[1] This passage is consistent with Lukacs in that it clearly repudiates what Marxists have stigmatized under the names of "voluntarism" and "spontanism." But Marx, unlike Lukacs, never thought of the working-class political party (or anything else for that matter) as an infallible authority on what the proletariat's interest is or what its class consciousness should be. Clearly for Marx these were issues open to free debate among members of the working-class movement and were to be decided on the basis of empirical evidence about what the proletariat's "historical being" is, and what that "being" necessitates the proletariat to do in order to emancipate itself.

Harrington does not clearly distinguish this point from the related but quite distinct (and equally correct) one that Marx (unlike some later Marxists, including Lenin) did not equate morality—even "Marxist" morality—with whatever conduct happens to further the interests of the proletariat. (This is true whether or not these interests are to be defined by some pontifically infallible party elite.) Marx never thought that an idea was correct just because it was a proletarian idea or because its acceptance would serve proletarian interests. On the contrary, proletarian interests are progressive interests at least in part because these interests cohere with ideas (such as Marx's social theory) that are objectively correct—or at least as objectively correct as the human mind is capable of coming

---

[1] *Karl Marx—Frederick Engels: Collected Works*, 9 vols. (New York: International Publishers Co., 1975–77), vol. 4: *Marx and Engels 1844–1845* (1975), p. 37.

up with at the present time. But not every idea that might prove advantageous to working-class agitators is true or well founded; whether it is so must depend on the evidence for it. In particular, Marx did not think that we can show moral norms to be valid simply by showing that their dissemination or satisfaction would serve proletarian interests. In the *Critique of the Gotha Program,* Marx insists that the bourgeois are *correct* when they assert that the present social system is distributively just, despite the obvious fact that it would be ideologically advantageous to the proletariat to claim that capitalist exploitation of the workers is unjust. (I will be returning to this point later.)

Second, I think Harrington is correct in saying that throughout his career Marx believed that human nature determines certain things as essential elements of human well-being. Marx was, in more familiar philosophical parlance, a naturalist. He thought that by studying human beings, finding out what sorts of organisms they are, and examining their normal functioning (both physical and spiritual), one could determine empirically that certain things are good for them or in their interest. Harrington emphasizes that human nature for Marx is not fixed or static but historical. Again, he is right. The activities and conditions that benefit human beings at one stage of historical development do not benefit them, or at least do not benefit them to the same degree or in the same way, at a different stage. Human needs change in history and on the whole expand. But none of this undermines the basic idea that what is good for people can be determined by an empirical examination of the kind of beings they are. We must only be careful to include historical factors in this examination.

Elizabeth Rapaport suggests that Marx's historical materialism was developed at least in part to justify the conception of the human good that Marx had earlier held on more naive, Feuerbachian grounds. I think she is right. To tell the whole story here would take a long time. Briefly, I think that what is distinctive about Marx's conception of the human good is that it is focused on human social self-expression in labor and on the historical growth of humanity's laboring capacities. Marx's historical materialism provides support for this conception by showing (if, as a theory of history, it proves empirically successful) that the basic intelligibility of human history is provided by humanity's Promethean expansion of its productive powers. Social structure and social change, on the materialist theory, are to be explained by the manner in which social forms accommodate the existing productive powers of society and the way in which they contribute to the further development of these powers. This theory does lend support to the idea that productive development is the most basic human aspiration, the fundamental drive that moves peoples, societies, epochs. This does not strictly entail but it does strongly suggest that the open-ended development of productive powers is the fundamental characteristic of normal human functioning and hence the fundamental human good, or the focal point socially and historically for all human goods. Marx never makes fully explicit this motivation for advancing historical materialism. But I think it is one of the things that motivates his view and that Rapaport has shown considerable insight in bringing this to our attention.

Harrington's paper gives the impression that human well-being for Marx can be summed up in a single value: freedom. I wish he had said more about what this term is supposed to mean. Marx was a explicit supporter of the notion of "positive freedom": freedom "in the materialistic sense," he says, is "not the negative power to avoid this or that" but the "positive might of making our true individuality count."[2] Freedom in this "positive sense" clearly covers more elements of human well-being (and is in itself more easily defensible as one of those elements) than the "negative freedom" championed by liberals and social libertarians. But positive freedom is only a determinate human good insofar as it is identified with rational self-determination. And this implies that there are other human goods, other elements of human well-being, that are there to be chosen by the

[2] Ibid., p. 131.

rational, self-determined agent. For this reason, I would resist the idea that freedom, even freedom in the positive sense, is the sole human good or the sum of human goods. In Marx's own formulation, "the positive might of making our true individuality count" the "might" to do this is distinct from the fulfilled individuality that is to be achieved through it.

In any event, a good case can be made from the Marxian texts for ascribing to Marx other things as important elements of the human good: "self-actualization" or "development" in Marx's jargon, the "development," "exercise," "actualization," "objectification," and "confirmation" of "the human essence" or of our "human essential powers" and community, an economic order in which people consciously and directly affirm and produce the life of others instead of being compelled by the social forms of private property and commodity production to treat their products only as means of gaining dominion over others. Beyond "freedom," "community," and "self-actualization," the list of elements of human well-being to which Marx subscribes could be extended to others, less philosophically exciting but no less important: physical health and comfort, security, enough food to eat. Marx attacked capitalism in the name of all these goods. It may not be farfetched to ascribe to Marx a "conception of human well-being," but it is misleading to suggest that this conception can be captured in a single value such as "freedom."

A third point: Harrington is wholly right in repudiating the truly asinine idea (espoused by such people as Karl Popper) that Marx inferred the desirability or progressiveness of a social development from its historical inevitability. In fact, the very reverse would be much closer to the truth. Marx tends to regard a social development as inevitable whenever it seems to be required by basic human needs and aspirations—by the need to expand society's productive powers or to actualize the full range of human potentialities these powers put within people's reach. Marx saw the rise of capitalism as inevitable in western Europe during the early modern period because—given the existing stage of productive powers and the precapitalist social conditions then and there prevailing—only capitalist relations had the capacity to concentrate and organize social production in the way modern industry requires. The overthrow of capitalism is equally inevitable because capitalism is incapable of putting these powers of production at the disposal of human beings. It is because people will eventually appreciate this fact and act on it that the downfall of capitalism is historically necessary.

Marx did see the rise of capitalism in western Europe as a progressive and historically inevitable movement because through it humanity acquired awesome productive powers and expanded its powers at an unprecedented rate. In his account of the grisly history of capitalism's victory over earlier, more idyllic social relations, his horror at its egregious inhumanity is mixed with a certain awe, even admiration, at capitalism's historical power and its very real accomplishments in humanity's behalf. Harrington perhaps does not emphasize this enough. But it is important: for Marx, a movement may be inevitable, progressive, and even in some respects admirable while at the same time being terrible, inhuman, and in other respects revolting and disgusting. But of course Harrington is quite correct to point out that Marx never saw the accomplishments of capitalism as any reason for tolerating its existence for an instant once there emerges the historical possibility of developing humanity's powers at less human cost.

The fourth and very crucial point Harrington notices is that what distinguishes Marx from other social critics (both before and since) is not his social "ideals" but his attempt to identify with and support a progressive historical movement, a movement that (on Marx's theory) can be identified as progressive even in the absence of any but the vaguest notion of the sort of society it will bring about. Any honest social reformer, Marx told Ruge in 1843, "must admit to himself that he has no exact view about what ought to be. But this is just the advantage of the new trend, that we do not dogmatically anticipate the world, but only want to find the new world through a critique of the old one."[3] I would even go so far as to say that Marx denied having any "ideals" at all. At any rate, Marx himself went this far: he denied that communism is an "ideal" or a "state of affairs which ought to be brought about." Rather, communism is for Marx "an actual movement which is abolishing the present state of affairs."[4] "The workers," he said in the Civil War in France (1871), "have no ideals to realize." Their task is only "to posit freely the elements of the new society" in the course of "long struggles, historic processes transforming circumstances and men."[5]

Harrington does speak of Marx's "ideals." I think this involves no distortion as long as we keep clearly in mind that Marx had no intention (in his own contemptuous phrase) to "concoct recipes for the cookshops of the future."[6] The point to be emphasized—which Harrington plainly sees—is that for Marx revolutionary practice consists not in formulating goals and then trying to realize them but rather in identifying, cultivating, and supporting a progressive social movement already in existence, whose historic mission and whose potency to carry it out are based on historical circumstances and human needs that are quite independent of any "ideals" reformers may devise for themselves at various stages of the movement. Marx clearly believed that one can identify such a movement as progressive while having only vague (largely negative and always modifiable) beliefs about the kind of society it will create. The basis of all Marxian practice is the actual existence of some such movement of this kind. If a socialist were to surrender the belief that there is a progressive, proletarian movement actually in existence, then it seems to me that he

[3] *Karl Marx—Frederick Engels: Collected Works*, vol. 3: *Marx and Engels 1843–1844* (New York: International Publishers Co., 1975), p. 142.

[4] *Karl Marx—Frederick Engels: Collected Works*, vol. 5: *Marx and Engels 1845–1847* (New York: International Publishers Co., 1976), p. 49.

[5] *Selected Works of Marx and Engels* (New York: International Publishers Co., 1968), pp. 294–95.

[6] Karl Marx, *Capital*, 3 vols, vol. 1, (New York: International Publishers Co., 1967), p. 17.

would have *eo ipso* surrendered his claim to be a Marxist, whatever other beliefs he might share with Marx and however much he might sincerely praise Marx's work and respect it. It might be a nice question to decide how far Harrington in his various writings has really abandoned this basic tenet of Marxism. But it is not a question I will try to decide here. In any case, the problem of how far a socialist who abandoned Marxism in this basic way would have fallen into error would be a separate question altogether. Perhaps he would only have recognized an obvious historical truth about the twentieth century. This is also a question I will not try to decide here.

I turn now to the main point where (I think) I disagree with Harrington's interpretation of Marx. Harrington insists that Marx's consistent and sometimes savage attacks on moralizing social criticism are not to be read as hostility to morality itself or to the use of moral ideas in politics. I find it difficult to read them in any other way. Marx was not averse to morality in evaluating people's conduct in private life, as Harrington notes. Marx often attacks individuals for such political vices as opportunism, mendacity, and self-serving collaboration with the enemy. He praises the "self-sacrificing heroism" of the Paris communards and scorns the French bourgeoisie of 1848 for sacrificing its own class interest to the "filthiest private interests."[7] But Marx does show a consistent aversion to using distinctively moral norms, such as "right" and "justice," in the assessment of fundamental social relations. Capitalism for Marx is bad, but not *morally* bad. In a few passages, Marx even seems to think that his world view has in some sense undermined morality as a whole. In *The German Ideology* he declares that historical materialism has "broken the staff of all morality" by showing the foundation of all moral ideas to lie in the class interests they represent.[8] The *Communist Manifesto* says that to enlightened proletarians, the dictates of morality are "so

---

[7] *Selected Works of Marx and Engels,* pp. 305, 160.
[8] *Karl Marx—Frederick Engels: Collected Works,* vol. 5, p. 419.

many bourgeois prejudices, behind which lurk just as many bourgeois interests."[9] A bit later, the *Manifesto* responds to an imaginary critic who charges that "communism does away with morality instead of founding it anew." The response does not deny that the charge is *true* but only observes that the communist revolution, by making the most radical break with traditional property relations, must involve a most radical break with traditional ideas as well.[10] The clear implication is that "doing away with morality instead of founding it anew" is part of this radical break.

I am willing to admit that in these flashy, iconoclastic passages Marx is overstating his real views. At least, I see no way in which a wholesale repudiation of morality can be rendered consistent with Marx's own moral judgments about individuals, which are scattered through his writings. Yet I do think we must take Marx's repudiation of morality seriously. In particular, Marx does consistently repudiate moral norms (such as right and justice) as vehicles of radical social criticism. He does so, I believe, because he sees morality as serving a definite social function: that of motivating people (largely through ideological illusions and superstitions) to fulfill the functions required of them by a social order or class movement. Most fundamentally, the social function of morality is to sanction the smooth functioning of a mode of production. Thus, in *Capital,* Marx asserts that the justice of transactions between agents of production rests on the way these transactions correspond to the prevailing mode of production.[11] A transaction is just if it corresponds to the prevailing productive mode and unjust if it contradicts that mode. Quite consistently with this, Marx argues in a number of passages that although capital exploits labor, it does the workers no injustice and does not violate any of their rights but is entitled with full right to

[9] *Karl Marx—Frederick Engels: Collected Works,* vol. 6: *Marx and Engels 1845–1848* (New York: International Publishers Co., 1976), pp. 494–95.
[10] Ibid., p. 504.
[11] Marx, *Capital,* vol. 3, pp. 339–40.

what it squeezes out of its victims.[12] This is consistent with Marx's views, because he holds that the exploitation of labor by capital is essential to the capitalist mode of production, not a mere abuse carried on within it. Exploitative transactions between capital and labor *correspond* to the capitalist mode of production and hence they are just as long as that mode of production prevails.

Needless to say, Marx does not regard the justice of capitalist exploitation as any defense of it. Only a person who has not yet seen how morality is by its very nature a cloak for ruling-class interests could think otherwise. Whenever would-be champions of proletarian emancipation state their demands in moralistic terms (as did the Gotha Program) Marx's response is to agree (at least verbally) with bourgeois apologists and to *deny* that the moral charges such critics bring against capitalism are true. The "proletarian" moral standards by which such critics want to condemn capitalism are wrong standards, at least standards inapplicable to capitalism.

For Marx the basic social function of morality is a conservative one. In holding this view, Marx was following Hegel, who regarded the basic social function of morality as that of embodying and preserving a spirit, a people, a culture, a way of life. In the progressive movement of history, Hegel declares, "there arise great collisions of subsisting, recognized duties, laws and rights with those possibilities which are opposed to the system, which violate it and even destroy its foundations and actuality."[13] These possibilities are actualized in Hegel's view by those "world historical individuals" who respect none of the limitations which laws and morality would impose on them. Historical progress is always the triumph of moral evil.

[12] See *Selected Works of Marx and Engels,* pp. 321–22; and Marx, *Capital,* vol. 1, pp. 194, 584–85.

[13] G. W. F. Hegel, *Lectures on the Philosophy of World History: Introduction,* trans. H. B. Nisbet (Cambridge, England: Cambridge University Press, 1975), p. 82.

Marx and Engels were aware of these views in Hegel and explicitly endorsed them.[14] Of course, Marx never thought of history as made by "world historical individuals" or "great men." But it *is* made by revolutionary classes. And a revolutionary class is, according to his account, a ruthless destroyer of everything recognized as holy and virtuous, right and just. The bourgeoisie treated feudal, guild, and petty industrial mores in this way; when the proletariat emancipates itself, it too must do evil and injustice. The *Manifesto* openly advocates "despotic encroachments on property rights" to this end.[15]

Of course the moral standards violated in a revolution will be the standards of the old order. The property rights violated by a proletarian revolution will be rights whose validity consists in the way they correspond to the capitalist mode of production and thus serve bourgeois interests. But in Marx's view this is what makes them the *correct* standards. If we think there are other, better, or truer standards of right or justice by which capitalism should be measured, then we are thinking about morality in a way that is fundamentally at odds with Marx's way.

Naturally once the proletariat has introduced a new, socialized mode of production, different standards of right (which Marx adumbrates in the *Critique of the Gotha Program*) will come to prevail, and transactions that are exploitative will count as unjust transactions. The point not to miss, however, is that these new standards of justice will come to apply only when and only because a revolution has taken place and a new mode of production has come into being. It is *not* Marx's view that a social revolution does occur or should occur because socialist moral standards are already valid for capitalist society. To think this is (in the words of the *Critique of the Gotha Program*) to

---

[14] See *Selected Works of Marx and Engels*, p. 615; and *Karl Marx—Frederick Engels: Collected Works*, vol. 6, p. 174.

[15] *Karl Marx—Frederick Engels: Collected Works*, vol. 6, p. 504.

suppose that juridical concepts rule economic relations, where the truth is that juridical relations arise out of economic ones.[16]

Earlier I agreed with Harrington that Marx valued freedom, and wanted to overthrow capitalism so that people might be emancipated. I even added community and self-actualization, among others, to the elements of human welfare for whose sake Marx sought to overthrow capitalism. But are these not moral values? My answer is that they are not. Of course there is a bland and inclusive sense of "moral" in which any good whatever is a "moral" good and any view about what is good is a "moral view" or even (to put it pretentiously) a "moral theory." But there is a narrower and (I think) more proper sense of "moral" in which moral goods can be distinguished from nonmoral ones. What is morally good is what conscience or the "moral law" tells us to do, or to be, or to aim at,—it is what we are apt to be blamed for not doing or to feel guilty for not doing. Moral goods include such things as acting justly, doing our duty, respecting others' rights, having and displaying a good will, and cultivating a virtuous character. Nonmoral goods, on the other hand, are things we would want (for ourselves or for others) quite apart from considerations of conscience or duty or guilt or love of virtue. They include pleasure and happiness and (I submit) all the goods for whose sake Marx wanted capitalism to be overthrown: freedom, community, self-actualization, prosperity, security, bodily health.

No doubt some moralities enjoin and sanction the pursuit of these nonmoral goods. But that does not turn them into moral goods. The crucial point is that Marx never appeals to moral considerations in urging people to overthrow capitalism and achieve these nonmoral goods. He never claims that the workers should be emancipated because they have a right to be free, or that we should work for a more human society because virtue or duty bids us to. Marx evidently felt that the obvious nonmoral value of freedom, self-actualization, and other such goods is by itself sufficient (without appeals to our sense

---

[16] *Selected Works of Marx and Engels*, p. 322.

of guilt or love of virtue) to convince any health-minded person to favor a social system that achieves them and to oppose a social system that needlessly deprives people of them. The same considerations explain why Marx was no friend to "proletarian" moral ideologies that demanded an end to capitalism on the grounds of its alleged "injustices." Such ideologies, like all moral ideologies, are only a mask for class interests. But one of the chief ends of the proletarian movement, as Marx conceives it, is to emancipate people from all such mystifications. Proletarian demands phrased in moral terms may have corresponded to an earlier, less mature form of the movement, before historical materialism had clarified the real nature of the historical process. But to present these demands in moralistic terms is now mystifying, obfuscatory, and retrogressive.[17] The correct course is to point to the empirical facts: that capitalism needlessly frustrates important nonmoral goods which socialism would begin to actualize. (These facts are amoral, but by no means "value-free.")

Marx's writings seethe with anger and indignation against the complacency and hypocrisy of the bourgeoisie and its academic sycophants in the face of capitalism's inhumanity. But there is no inconsistency in being morally indignant at complacency in the face of massive and remediable nonmoral evil while refusing to condemn morally the nonmoral evil itself. And while Marx may express and provoke anger, he never tries to make anyone feel guilty for being a bourgeois or playing the role of exploiter. In the preface to *Capital,* Marx goes out of his way to point out that he does not hold capitalists and landowners morally responsible for social relations whose creatures they are, however much they may subjectively rise above them.[18] Marx has often been accused of being a hatemonger. Perhaps he was. But he was never a guiltmonger. As compared with the liberal guilt of bourgeois over the "social injustices" from which they benefit, proletarian class hatred is a clean, healthy, honest emotion.

[17] Ibid., p. 325.
[18] Marx, *Capital,* vol. 1, p. 10.

Marx never explicitly drew any distinction between moral and nonmoral goods. But the distinction is a familiar one both in philosophy and in everyday life. It is not implausible to think that Marx might have employed it without ever taking explicit note of it. If we do ascribe such a distinction to him, we can see as consistent passages that would otherwise be hopelessly inconsistent. For instance, when Marx claims that historical materialism "breaks the staff of morality" and that morality is only bourgeois prejudice masking bourgeois interest, we would not want to infer that he means that historical materialism undermines such values as freedom, community, and self-actualization, in whose name Marx attacks capitalism; we would not want to conclude that he is stigmatizing those values themselves as bourgeois prejudices.

When Harrington says that Marx is not hostile to morality, he may be using "morality" in the bland and inclusive sense that includes nonmoral values as well as moral values and has no special connection with justice, right, duty, guilt, virtue, and other such things. In that case, my disagreement with him would be more verbal than substantive. But in that case his claim that Marx is not hostile to morality would not say very much: it would say only that Marx attacks capitalism in the name of some values or other. (It is hard to see how anyone could attack anything as vehemently as Marx attacks capitalism without esteeming some values or other in whose name the attack can be made.) Further, there would still be a serious shortcoming in Harrington's discussion. For I think that Marx's consistent and open hostility to moral values (in the narrower, more proper sense), his adamant opposition to the employment of such values in the assessment of fundamental social institutions, is one of his most pervasive, striking, unconventional, and distinctively Marxian views about morality. Any exposition of "Marxist ethics" that ignores or sloughs over it has, I think, forfeited its claim to be an exposition of Marx's own views on such topics.

PART THREE

Freud

CHAPTER SEVEN

Freud's Impact on Modern Morality and Our World View

ROBERT R. HOLT

The world is in a moral crisis today, everyone agrees. It is little comfort that many people would have agreed with the preceding statement at just about any time in history. The older generation has worried about the decline in the moral fiber of the young at least since the time of the Greek philosophers, and we still have not quite succumbed. Such a general reassurance can hardly substitute for data, but it would be difficult indeed to conduct an empirical survey of the present moral state of the world, quite aside from the fact that comparable data from earlier times would be even more nearly impossible to obtain. In our huge and pluralistic world, there are many evidences of increased respect for life and for individual liberty, more social justice, and other moral desirables. It is very difficult to balance such trends against the many signs that monetary values override all others in many contexts, that unjustifiable exploitation and privilege are firmly entrenched, and that various types of crime are on the rise. Overriding everything else, however, is the unspeakable moral horror of our time, the declared readiness of the great powers to pursue their national interests by war with stockpiles of nuclear weapons

Preparation of this paper was supported by a United States Public Health Service Research Career Award, Grant No. 5-K06-MH-12455, from the National Institute of Mental Health.

that could permanently end all life on earth[1]. Clearly, the traditional moral systems that have been an indispensable prop
to society and civilization are failing to provide the most fundamental controls needed to enable life itself to go on.

Many observers have noted, with varying degrees of
uneasiness or alarm, that the secular trend of religious decline
is in the process of depriving moral values of their major source
of legitimacy and emotional leverage on mankind. I cannot
attempt to determine how far the alleged slackening of religious faith is related to many alarming evidences that a lack
of internalized moral controls threatens our future.

Nevertheless, these worrisome trends are important enough
to warrant our considering the often-heard arguments blaming
many of them on the psychological outlook in general and on
Freudian psychoanalysis in particular. I shall take up the charge
that psychoanalysis has directly and indirectly gnawed away
at our moral underpinnings as well as the counterclaim that
it offers a new and (many claim) superior mortality with a
rational, scientific base; I shall also try to suggest ways in
which the impact of psychoanalysis on morality should be
examined in a less passionate context than attack, defence,
and counterattack.

IS PSYCHOANALYSIS THE ENEMY OF MORALITY?

Freud considered himself an unusually moral man and
strongly resented accusations to the contrary. He saw these
unjustified attacks as arising from three main sources. First,
antagonized by his atheism and his refusal to adopt the conventional public silence of many other unbelieving scientists,
organized religion assumed that he must be not only ungodly
but immoral. For Freud insisted on pointing out the Christian
churches' hypocrisy and attacking their support of trends he
considered antithetical to mental health. Second, he rejected

[1] J. Schell, *The Fate of the Earth*, (New York: Avon, 1982).

certain suppressive, inhibitory rules of conventional morality, especially antagonism to sexual pleasure, which he believed were contributory causes of neurosis. Hence, it was easy to misperceive (or misrepresent) him as opposed to *all* morality. His opposition to the moral strictures he considered pathogenic took several forms; he diagnosed many "saintly" persons as mentally ill and attributed much neurosis to an overly severe superego. He developed a form of treatment aimed in part at changing patients' ethical standards, getting rid of the burden of sinfulness and self-destructive guilt, helping them to be more assertive of their own desires and more willing to express their impulses and enjoy a full sexual life. He encouraged patients to replace an automatic taboo morality with a more rational and voluntary self-control. Freud also advocated new patterns of childrearing, which emphasized more freedom to express impulses, more capacity for pleasure, and less guilt (but not license). To many a shocked late Victorian, these three applications of psychoanalysis seemed to undermine society's moral foundations.

Finally, and perhaps most fundamentally threatening, Freud supplied a new kind of ammunition to the anticlericals: he put religion and morality under scientific scrutiny. When as pious (though Protestant) a man as Kant had subjected man's moral feelings and behavior to intellectual analysis, his works were put on the Roman Catholic Church's Index. How much greater a threat was this impious attempt by an agnostic Jew to strip the conscience of its sacred nimbus and to view it as a potentially psychopathological force! Not only did he commit the sin of barging into the holy places of the human spirit, with microscope rather than hat in hand, but he insisted on finding there the evidences of unholy impulses actively at work and on tracing many of humanity's miseries to what had been considered the finest side of our nature, the citadel of the spiritual within each of us. Freud saw that the image of God is transparently the imago of the beloved but feared father of early childhood and that the rituals of his worship are closely akin to obsessional symptoms. Religion is "comparable to a childhood neurosis," he said, or even to a shared delusional system.[2]

Our moral scruples and restraints, Freud taught, are attributable to the operation of a "psychic instance," or some kind of structural agency, which he called the *Überich*. Though we are familiar with the latinized translation "superego," let us reflect that his coined German word implies an *Ich*, a sense of "I," and something felt to be over it, superior (in power, not in virture) to the core of selfhood and thus at least partly separate from it. The debate over what is meant by the term *superego*, how justified Freud was in endowing observed feelings and other reported inner events with substancelike properties,[3] and the scientific legitimacy of his entire "structural model"[4]—these matters interest me greatly (and I have strong and unorthodox views about them), but it would take me far afield to enter this highly controversial theoretical arena. At the moment all I wish to do is to remind you of Freud's insistence on treating every aspect of human thought, behavior, and feeling with the same scientific detachment. Nothing was too disgusting or too sacred to be analyzed; and if psychoanalysis was often reductionistic, it was because scientific method—as he understood it—made that necessary. Freud considered the unconscious wishes and fears, to which he traced much that was superficially innocent or lofty, far more interesting and important than conscious ideals, intentions, and other mental furnishings. He was all too inclined to say that guilt, for example, was *nothing but* blind self-punishment, laying aside entirely the question of how justified a man might

[2] Sigmund Freud, *The Future of an Illusion*, (1927) *Standard Edition*, Vol. 21 (London: Hogarth, 1964).

[3] Roy Schafer, *A New Language for Psychoanalysis* (New Haven: Yale University Press, 1976).

[4] R. R. Holt, "The Past and Future of Ego Psychology," *Psychoanalytic Quarterly* 44(1975):550–76.

be in his self-reproaches over, let us say, the abandonment of his family for a new infatuation.

Freud felt strongly that psychoanalysis would be made powerless as a method of scientific inquiry and of therapy if the analyst allowed himself to react with moral condemnation toward a patient. He makes it quite evident, in his clinical writings, that he had to suppress or hold in abeyance his own repugnance for deviant sexual practices or fantasies, his personal horror over incestuous or parricidal impulses and dreams, and the like, in order to be able to help the patient to confront these unpleasant and threatening sides of himself and to understand them. Freud has often been misunderstood on this point. He did not say or mean that "anything goes" or that the analyst should in any way encourage or condone most forms of socially deviant behavior, even in the realm of sexuality. It is a common observation by those who have actually read Freud that he was surprisingly prudish, very much the proper Victorian, not only in his own conventional and inhibited sexual life but in his evident reaction to such "degenerate" practices as oral sex. He never wavered in his conviction that masturbation causes the serious damage of "actual neurosis." Evidently, he himself coped with the personal problem of having to hear constant talk about sex, including declarations of love and attempts on the part of attractive patients to seduce him, by the defense of isolation and intellectualization.

Another, related aspect of Freud's personality had far-reaching implications for our topic. Reading the minutes of the Vienna Psychoanalytic Society or some of Freud's letters to other analysts, one may be shocked by his occasional tone or moralistic condemnation and denunciation. Freud was capable of excoriating some people he encountered clinically as "scoundrels," "absolute swine," "worthless," and "good for nothing."[5] He once wrote to Pfister:

[5] Paul Roazen, *Freud and His Followers* (New York: Knopf, 1975), pp. 145ff.

I do not break my head much about good and evil, but I have found little that is "good" about human beings on the whole. In my experience most of them are trash. . . . If we are to talk of ethics, I subscribe to a high ideal from which most of the human beings I have come across depart most lamentably.[6]

After reading such statements, one could hardly content that psychoanalysis necessarily eliminated moral judgments, converting the analyst's reaction toward lawbreaking behavior into a cool, dispassionate concern only for its roots in personal psychopathology!

The first question, about the direct and supposedly adverse impact of psychoanalysis on morality, must be answered on the whole in the negative. Surely for Freud himself, becoming a psychoanalyst did not alter his own starchy adherence to a conventional moral code. He did not intend his new discipline as a direct weapon against morality, though he was opposed to certain specific moral rules and to organized religion. He believed that a well-analyzed person would, if anything, be more ethical than before treatment because she or he should become more rational and flexible in applying more principles.

One may well object that though Freud may not have intended psychoanalysis as an attack on sexual morality, it did in fact become that. In the cheapened, vulgarized form in which psychoanalysis filtered down to the general public, it was misunderstood by many as advocating the complete dissolution of the superego and a libidinal utopia of universal licentiousness. Witness the following snatch from a popular song of the 1960s:

> Glory, glory psychotherapy,
> Glory, glory sexuality,
> Glory, glory, now we can be free,
> As the id goes marching on!

---

[6] E. L. Freud and H. Meng, eds., *Psychoanalysis and Faith: Dialogues with the Reverend Oskar Pfister,* trans. E. Mosbacher (London: Hogarth Press, 1963), pp. 61–62.

Can Freud be held responsible for Wilhelm Reich, Herbert Marcuse, and their even less intellectually disciplined followers and popularizers? Only in the sense that if he had not lived, they could not have thought and taught what they did. But any doctrine can be misunderstood, and almost any truly original idea tends to be disseminated to the public in a form that its author would repudiate with horror.

Psychoanalysis, Freud insisted, was for the worthy few. He handled the problem of curbing his scornful and rejecting reactions to patients who had multiple perversions or engaged in criminal behavior by classifying them as unanalyzable; therefore, he had only fleeting professional contact with them. That bit of self-protective behavior had further implications for our topic, which are often overlooked. Since he analyzed only people whom he considered "worthy"—basically decent and truthful citizens who shared many of his own values—it was much easier for him to tolerate their guilty fantasies and their occasional, restrained acting out. As a result of this kind of screening process, the great majority of his patients were probably overcontrolled people (as he was himself) whose problem with respect to moral constraints was *not* to get control over their unethical behavior but to learn to relax their too-severe self-discipline. Yet on the basis of this biased selection of persons, Freud developed a set of theories intended to be quite general and applicable to all people.

We have serious need of a valid psychological understanding of moral thinking and behavior, one that is not limited to any particular type of person but is very generally applicable. In today's world, there seems to be a shrinking minority of neurotically overcontrolled persons who need help in taming their savage superegos and a growing number of people who pursue immediate gratification in what they see as their own self-interest without regard for harm done to others.[7] I have

---

[7] They are, of course, often seriously mistaken in believing that they are looking out for themselves; let me only remind you of Hardin's "Tragedy of the Commons." (Cf. G. Hardin, "The Tragedy of the Commons," *Science* 162[1968]:1243–48.)

deliberately made that formulation wide enough to cover not only criminals, terrorists, addicts, and others who break rules for selfish reasons but also many who operate within the law in pursuit of profit above all other values, even those who in ignorance insist on the "right" to waste energy, reproduce freely, pollute their environments, kill other species, wage war, or otherwise endanger the lives of many. How can one not be concerned about morality when the survival of not only humankind but perhaps all life on earth is in real and growing danger, and when we know much more about what must be done to fend off such danger than about how to get people to make the necessary changes in their behavior! In large part, the needed changes amount to self-restraint, self-control, the postponing of immediate pleasures and the giving up of familiar, easy comforts—the very objectives of a great part of morality. From this perspective, morality is a device of social control, an indispensable primary way in which societies have always prevented the breakdown of social order and their own eventual self-destruction. Laws and law-enforcing agencies are at best supplementary devices; the basic job of *self*-restraint has to be done by citizens themselves. Therefore, we need the best theories available to help us first to understand how people come to control their socially destructive wishes, so we can find ways of helping them do a better job of it; otherwise we shall perish—if not in a sudden nuclear holocaust, then more slowly from heat death, mass starvation, or the destruction of our habitat.

THE PSYCHOANALYTIC EXPLANATION OF MORALITY

After this sobering look at the stakes, let us turn to Freud's explanation of moral behavior. One consequence of his restricted sample of human beings on which to found his theories was that he found morality so ingrained in them as to seem innate. His earliest formulations about normal control of impulse were predominantly cast in terms of biological evolution,

wherein he followed Lamarck and Haeckel.[8] Initially, he was
more impressed by the early onset of disgust and morality as
reactions to sexuality than by any polymorphic–perverse impulses of the young child; in discussions with his friend Fliess
he speculated that the regular appearance of these antisexual
reactions in early childhood was an instance of the so-called
biogenetic law: an inherited trace of an early phase in the
evolution of the species when man rose up on his hind legs
and got his nose away from its easy access to sexually stim-
ulating odors—a phase change now recapitulated in the development of the individual. As late as the *Three Essays on
Sexuality* (1905), he wrote:

It is during [the] period of total or only partial latency that are built
up the mental forces which are later to impede the course of the
sexual instinct and, like dams, restrict its flow—disgust, feelings of
shame and the claims of aesthetic and moral ideals. One gets an
impression from civilized children that the construction of these dams
is a product of education, and no doubt education has much to do
with it. But in reality this development is organically determined and
fixed by heredity, and it can occasionally occur without any help at
all from education. Education will not be trespassing beyond its appropriate domain if it limits itself to following the lines which have
already been laid down organically and to impressing them somewhat
more clearly and deeply.[9]

If his caseload had been that of a modern therapist in a child
guidance clinic or community mental health center located in
the socially disorganized center of a modern metropolis, it is
hard to imagine that he would have written such words! For
the most part, they have been quietly overlooked.

Contemporary psychoanalysts say nothing about any innate dams of morality and a great deal about the development
of the superego through identification. Freud's first approxi-
mation was the characteristically sweeping statement, "the

[8] F. J. Sulloway, *Freud, Biologist of the Mind* (New York: Basic Books, 1979).
[9] *Three Essays on Sexuality* (1905), *Standard Edition*, 7:177.

superego is the heir of the Oedipus complex."[10] That is, he understood the typical course of male development as being that when the little boy comes to realize that he simply cannot displace his father and possess his mother sexually, on pain of castration, he retreats to a *fantasy* of taking his father's place by being one with his father. In the service of that magical plot, he takes on many of the older man's qualities, including the giving of moral commands and prohibitions—to himself. Otherwise put, he masters the painful experience of having to submit passively to such strictures by actively embracing the role of the father and giving the orders that he himself must follow. Nowadays, analysts have found, often enough, that such fantasies pertain to parents of *both* sexes in both boys and girls, and that many of them can be traced back to preoedipal beginnings; but those are relatively minor corrections of a great insight.

Freud and his followers have assumed, mostly without taking any particular notice of it, that "the parents" adhere to a relatively uniform type of Judaeo–Christian morality, which was officially standard around the turn of the twentieth century. It was more or less centered on the Mosaic Ten Commandments as reinterpreted for a growing entrepreneurial society; thus, it included much of what (following Max Weber) we call the Protestant Ethic: prudent delay of immediate gratification in the service of eventual rewards—pie in the sky, or (more likely) wealth in later years from the shrewd investment of savings obtained by not spending all one's present income. Few analysts have speculated about what happens when the parents themselves do not have faith in these ideals, when the hope of heavenly reward and fear of eventual reprisal from God is lost, and when the dominant ethos emphasizes consumption rather than saving.

Instead, Freud and his epigones concerned themselves with what should be substituted for the automatic inhibitory control of an ego-alien, infantile superego in the process of

[10] *The Ego and the Id* (1923), *Standard Edition,* 19.

therapeutic change. There Freud showed himself a heritor of the Enlightenment: he put his faith in reason and science. His goal was, of course, not to turn timid celibates into libertines but to help the patient bring his innate drives under rational control. Freed from the blind necessity to act out infantile fantasies of the dire consequences that would follow if he or she were to assert the normal perogatives of an adult, the cured patient would settle issues on their merits. Or so Freud thought, again implicitly assuming a deep enough commitment to a set of common moral ideals so that the ex-analysand would be silently guided toward good rather than evil means of gaining realistic gratification of wishes. Perhaps he really failed to see that life is filled with legitimate moral dilemmas—situations in which one cannot simultaneously attain more than one ideal end and where logic or scientific information do not tell us which way to go. He expressed a truly touching faith that the soft voice of the intellect would prevail over what he was fond of portraying as the demonic forces of the id; but he did so partly because he still believed that the force of evolutionary biology supported this course.[11] In *The Future of an Illusion*, he argues first that the prohibition against killing need not rely on a faith that God has forbidden it, since it is possible to see how the fabric of society would fall apart if people were to kill one another opportunistically whenever they could get away with it. But then he harks back to *Totem and Taboo* and reiterates his faith in the myth of the primal horde, the notion that the contemporary horror of killing is our biological memory of the guilt felt by the original brothers after the slaying of the primeval father.

One of the many assimilative misunderstandings of Freud is to portray him as having the rather sociological view that society imposes its moral values on the child through the parents' socializing efforts. While it is possible to amass many quotations that seem quite consistent with such a root proposition, it is subtly but decisively *not* the way he thought about

---

[11] *The Future of an Illusion* (1927), *Standard Edition*, 21:53.

the issues. Being largely untouched by sociology altogether, he had no clear conception of society beyond that of common sense. In his most social-psychological works, the emphasis throughout is entirely on the standpoint of the individual. Nor did Freud have any inkling, in his latter years, of what was happening in anthropology, when the functional school came into being. The idea that a culture could be regarded as having a significant structure (much less one related to its institutional organization), even that it constituted a major part of his patients' psychic reality, was foreign to him. He knew anthropology only in its early evolutionary guise, as a collection of interesting travelers' anecdotes, which might be assembled by a Frazer into an esthetically appealing though artificial evolutionary progression. Therefore, he had to approach the task of explaining morality theoretically unaided (or should one say, unencumbered?) by any factual details about the relations between specific systems of moral values and the rest of their sociocultural contexts.

Being thus ignorant, Freud could have no idea about ways a person might be influenced, quite unawares, by cultural traditions or traits; it had to be left largely to others to apply and extend in this way Freud's idea of unconscious determination of thought, feeling, and behavior. He himself felt that social psychology could make no progress whatever without the Lamarckian assumption that cultural experience was unconsciously transmitted via the germ plasm,[12] and he assumed that after many generations of experiencing external restraint on impulses, people would have inherited a modicum of innate morality. Further, he thought that the process of learning parental moral ideals by identification was to a substantial degree guided by innate residues of the primal-horde experience. For example, it is not necessary for fathers to threaten to cut off their sons' penises at the height of the Oedipal conflict for the boys to develop a conviction that they will be castrated if they persist in sexual rivalry; the racial heritage of the primal horde

[12] *Moses and Monotheism* (1939), *Standard Edition*, 23:100.

guarantees that castration anxiety will naturally develop and aid the process of renouncing incestuous and parricidal wishes. To be sure, parents can help somewhat by carefully avoiding any such crude threats, but Freud did not believe that it would be possible, through good parental behavior, to prevent the castration complex entirely.

Whatever aspect of morality Freud considered, he characteristically reduced it to an internal problem of controlling unruly wishes. In considering the highest and most admirable forms of ethical, artistic, and scientific thought and behavior, he conceptualized the central process as one of sublimation— a transformation of the hypothetical energy that powered primitive, mostly sexual desires into a restrained, more manageable form. Likewise, in attempting to understand criminal behavior, Freud put forward the useful insight that there are many criminals who break laws out of a sense of unconscious guilt and allow themselves to be caught so as to bring on external punishment. In doing so, they are acting out an unconscious scenario that demands punishment in the denouement, albeit for some infantile sin such as masturbation. It is true, I believe, that some people do act in just such ways, and that they can be helped to get out of a repeated pattern—as self-defeating as any other neurosis but in this case socially destructive as well—by psychoanalytically informed treatment. Notice, however, that Freud tended to recast the entire social problem of crime as one of personal impulse management.

Any valid contribution to knowledge is, in my view, a positive achievement. In several ways, Freud and his followers have given us important insights into many aspects of immoral and moral behavior alike. At the same time, Freud's reductionism equivocated his achievement. His attempt to *explain* his valid clinical observations by recourse to a curious abstract theorizing he called metapsychology had two reductionistic aspects: First, the fundamental concepts of metapsychology (forces, energies, and structures) were drawn from physics and reflect Freud's conviction that ultimately psychology would be replaced by biology, which in turn could be reduced to

physics and chemistry. Second, Freud had a strong tendency
to formulate an explanation by saying that a phenomenon was
nothing but an expression of an unconscious tendency. In that
same spirit, he was convinced that because psychoanalysis
had unique access to unconscious wishes and defenses, which
often seemed to override all conscious and rational considerations so dramatically, his discipline alone had the *real answers* to human problems. Enthusiastic partisans for new ideas
often overvalue them in just this way, and no doubt the early
psychoanalysts' strident and even patronizing rejection of
everyone else's ideas as "superficial" can be attributed in part
to their exposed position. For political reasons, they exaggerated the degree to which they were lonely pioneers, bearing
aloft the banner of unwelcome truth despite universal attack
and rejection; but there *was* a kernel of truth in it.

Unhappily, the counterrejective stance persisted even when
the psychoanalysts achieved their moment of triumph, in the
two decades around the middle of this century. They still tended
to act imperialistically, claiming a profounder insight into criminal behavior, for example, than that of nonpsychoanalytic
criminologists, sociologists, anthropologists, and psychologists. The result was to make it difficult to integrate the contributions of all these disciplines. As a result, psychoanalysts
all too often went their own way, not realizing that their isolation was by now clearly self-imposed.

SOME INDIRECT EFFECTS OF PSYCHOANALYSIS
ON CONTEMPORARY MORALITY

I will turn now to a more speculative endeavor to develop
some hypotheses about subtler ways in which Freud and his
ideas may have affected morality—both theoretically (e.g., in
metaethics) and in the real world of good and evil.

Let me initially continue with the issue of explaining immoral and criminal behavior. Simple people often feel suspicious of any attempt to *understand* why others do wicked
things instead of directly condemning and punishing them.

They feel, perhaps with some justification, that once we suspend the punitive attitude of moral indignation, approaching a socially destructive person objectively yet empathically enough to come to an understanding of why he did what he did, we may end up sympathizing, excusing, and defending him against retribution. In a real world of a highly imperfect system of criminal justice, the scientific student does fairly often come to be an advocate for the captured criminal. People whose own moral development has not advanced to a postconventional stage (see below) often condemn psychoanalysts–and others who have been persuaded by psychoanalytic explanations of criminal behavior—as soft-headed bleeding hearts who are taken in and manipulated by evildoers. In this respect, however, the position of psychoanalysts hardly differs from that of the equally scorned sociologists or nonpsychoanalytic psychologists who also attempt to construct a theory instead of clamoring for the death penalty.

Psychoanalysts impinge on the criminal justice system in another way when they act as expert psychiatric witnesses. I trust that I may merely remind you of the sad and embarrassing spectacle that often ensues when two expert witnesses with equally impressive credentials speak out in court on opposite sides of a seemingly technical issue. The root problem here is that the legal system is predicated on an archaic, basically religious notion of free will, and psychoanalysis is like most of the other sciences in assuming a kind of determinism in which there is thought to be no possibility of freedom. The prescientific theory of behavior underlying the law assumes an immortal soul unconstrained by the causal forces of the material universe and therefore free to choose good or evil; therefore, the person is morally responsible for his acts unless it can be shown that his freedom was in some way artificially impaired. The various legal doctrines of limited responsibility come down to a determination of whether the person knows right from wrong, whether he was the victim of an irresistible impulse, or whether his free choice is impaired by mental illness—all matters on which psychiatrists are typically considered to be expert. In their own training, nothing is typically said about the soul, and free will is mentioned only in the process of their being told that Freud postulated an incompatibly exceptionless determinism, usually presented as one of his great legacies to modern psychiatry and psychology. Not only behavior but thoughts, fantasies, and even dreams are rigorously determined and thus are capable of being understood by the application of scientific laws. Hence, our task as scientists is not to condemn but to understand, not to hold a person morally accountable but to help him change toward more socially desirable ways of behaving. The incidental result of this argument is to undermine the rational basis for holding *anyone* responsible for his behavior without, at the same time, supplying a reasonably effective means of restraining antisocial acts.

Few issues in philosophy have piled up more volumes of disputation than that of free will. I cannot hope to make my position wholly clear, much less persuasive, without devoting much more time to it than is available here.[13] Briefly, I agree with those who argue that freedom of choice does not necessarily presuppose a dualistic or supernatural soul and is not at all incompatible with determinism as understood in contemporary science.[14] Hence, there is a sound basis for holding people accountable for their behavior and its consequences and a need to restructure the laws accordingly. But I think it would be hard to convict Freud of being very largely responsible for the general prevalence of disbelief in the freedom of the will among scientists and the educated elite of the Western world who have been much influenced by science. It is part

[13] Cf. R. R. Holt, "On Freedom, Autonomy, and the Redirection of Psychoanalytic Theory: A Rejoinder," *International Journal of Psychiatry* 3(1967):524–36; and "Freud's Mechanistic and Humanistic Image of Man," in *Psychoanalysis and Contemporary Science*, Vol. 1, ed. R. R. Holt and E. Peterfreund (New York: Macmillan, 1972), 3–24.

[14] Paul Weiss, "The Living System: Determination Stratified," in *Beyond Reductionism*, ed. A. Koestler and J. Smithies (New York: Macmillan, 1969).

of the philosophical undergirding of the sciences, the mechanistic world hypothesis that developed rapidly in the seventeenth century and reached a high-water mark in the late nineteenth century (see below).

It is not fair to Freud to leave the topic of responsibility without acknowledging that he was not inclined to take an easy way out, even at the cost of inconsistency. He raised the question: Must one assume responsibility for the content of one's dreams? He answered:

Obviously one must hold oneself responsible for the evil impulses of one's dreams. What else is one to do with them? . . . If I seek to classify the impulses that are present in me according to social standards into good and bad, I must assume responsibility for both sorts; and if, in defence, I say that what is unknown, unconscious and repressed in me is not my "ego" (*Ich*), then I shall not be basing my position upon psycho-analysis, I shall not have accepted its conclusions. . . . I shall perhaps learn that what I am disavowing not only "is" in me but sometimes "acts" from out of me as well.[15]

The psychoanalyst's job, then, was to help the patient overcome defensive efforts to disown his or her own wishes, even those that feel "ego-alien" and are outside voluntary control. Schafer has based his proposed replacement for metapsychology and his recommendations for treatment upon essentially these ideas—everything a person does is his action, however vigorously he may seek to disclaim it.

At the end of the paper just cited, Freud added:

The physician will leave it to the jurist to construct for social purposes a responsibility that is artificially limited to the metapsychological ego. It is notorious that the greatest difficulties are encountered by the attempts to derive from such a construction practical consequences which are not in contradiction to human feelings.[16]

[15] "Some Additional Notes on Dream-Interpretation as a Whole" (1925), *Standard Edition,* 19:133.
[16] Ibid., p. 134.

In this somewhat obscure passage, I think he is saying that the law needs to take a different position, holding a person legally responsible only for acts initiated and executed by his ego—by which I believe he meant what we have in mind when speaking of freely willed actions. His allusion to "contradiction" seems to me somewhat projective, for surely his position was in sharp contradiction to his own general doctrine of an exceptionless psychic determinism that precludes free will. Fortunately, whenever the practical realities of treating people in the real world clashed with his theoretical preconceptions, Freud had the saving grace to do what was clinically sensible and not worry about consistency.

Here we must distinguish two issues: First, to what extent can Freud and his movement be held responsible for the decline of religion? Second, how far is the decline of religion responsible for what is often called modern moral decay? I must admit immediately that it is quite beyond my competence to give definitive answers to these questions. To the best of my understanding, however, the hold of religion on the people of the Western world began to weaken before Freud's time, and students of the matter have been able to bring forward a great variety of plausible theories. Almost none of them can be empirically tested, unfortunately. Nevertheless, it is possible to adduce evidence that, in many countries, the shrinkage in church membership and in the evident power of religious ideas has had virtually nothing to do with psychoanalysis—for example, in the large part of the world dominated by communist ideology. Even where Freud has been widely read, I doubt that *The Future of an Illusion* has reached and influenced more people than has any other antireligious book. Within his own movement, there are a good many staunchly Freudian psychoanalysts who are also devout believers in God or are at least outwardly pious members of some religious congregation. They do not find it any more necessary to follow Freud in his irreligious stand than in his Lamarckism or his belief in a death instinct; psychoanalytic core ideas and professional practices do not require it.

Nevertheless, both Freud's direct attacks on religion and

the indirect effect of his scientific analyses of religious beliefs and rituals must surely have contributed in *some* part to the secularization of modern culture. How far is that trend responsible for the growth of various kinds of immorality? The consensus of philosophers since Plato, with the principal exception of the medieval moral naturalists, has been that there is no logically necessary relation between morality and belief in God (or adherence to any particular religious dogma). God wants us to obey His commandments because they are right; it is not that we know they ought to be obeyed (or that they are good) only because He wants it. The connection seems, instead, to be a more empirical one. Religious institutions are the main agencies concerned with transmitting, upholding, and reinforcing the traditional ethical norms of our culture. Moreover, religious institutions and universities are the only places where people may make careers of thinking about ethical issues, even possibly generating a morality that may better serve the needs of the present and future.[17]

Undoubtedly, it is in the family that children develop guilt, shame, and more desirable forms of deontic emotion—in short, their consciences—but the moral principles that parents teach and more indirectly inculcate almost always come from religious sources. Church and temple constantly support parents and urge them on in their socializing and moralizing efforts. Moreover, it is certainly possible that the family has weakened as an institution, especially in this socially controlling function, because of the decline of the church or at least for the same reasons. For example, the divorce rate in Europe and America has soared to heights hardly dreamed of in Freud's youth, as the religiously inspired taboo on breaking up the family "merely" for the personal happiness of the parents has diminished. Not without some justification, psychoanalysis has been accused of directly and indirectly accelerating the rate of divorces. I

---

[17] I want to take at least brief notice of the fact that a case can be made for the proposition that the influence of organized religion has *retarded* the moral development of humankind.

would not want to defend the proposition that merely holding unwilling and unloving partners together in an indissoluble marriage would have any positive effect on upholding morality, granted the rest of our cultural situation. Nevertheless, as we try to understand the present mixed moral picture, I think it only reasonable to allot some modicum of causal influence to psychoanalysis through its contributions to the weakening of religion and the family.

Another kind of indirect influence of psychoanalytic thinking must now be considered. One might argue that Freud's most revolutionary moral message was to call insistently and steadily, in Goethe's last words, for "More light!" In the corrupt and decadent society that supported him, he stood against self-deception and fought it in as many of its forms as he could, with what Philip Rieff calls his ethic of honesty.[18] In many ways, he believe the biblical assurance, "Ye shall know the truth, and the truth shall make you free" (John 8:32). To be cured, his patients had to become conscious that they themselves had wishes and fantasies which, though unconscious and in stark contrast with their conscious values, were guiding much of their behavior. Such medicine was particularly bitter for anyone brought up in the moral tradition that one must not even think evil thoughts. Freud's rejoinder to that stricture was succinctly stated by his young patient by proxy, Little Hans: "Wanting's not doing." Not only is it all right to *want* to commit incest and murder, it is necessary for your mental health to acknowledge that you—like everyone else—had fierce desires of these crude kinds as a child and have not gotten over them merely by having put them out of mind. That was the inexorable requirement of Freud's theoretical understanding, and it was backed up by many therapeutic successes.

In that respect, Freud demanded of his patients a change of moral standards that our popular culture is far from ready to adopt. Witness the tremendous flap over Jimmy Carter's admission, in the famous *Playboy* interview, that he had lusted

---

[18] Philip Rieff, *Freud: The Mind of the Moralist* (New York: Viking, 1959).

after women other than his wife. The older morality assumed that very thought itself was evil, so zealous was it to control behavior and so little did it trust people not to act out whatever they could conceive. We should not be too hasty in assuming that Freud was right. It is a real, empirical question whether psychoanalyzed people do restrain themselves from immoral acts as well as those who succeed in continuously repressing the corresponding wishes.

There exist a few communities where a high proportion of the members have been analyzed—Vienna (the circle around Freud); Topeka, Kansas; Stockbridge, Massachusetts; and probably several other such psychoanalytic centers—not the whole city or town, but the enclave of psychoanalytically trained people, living and working together more closely than is possible in, say, New York. Casual observation supplemented by gossip is all that I have by way of data on what goes on in these centers, but that much is not encouraging to the hope that people in these groups would behave in a *generally* better, more ethical way than people in other strata of society. Rather, it is likely that such people are law-abiding but also good at acting individualistically and rationalizing minor moral lapses, and that they are relatively unrestrained by guilt in their search for self-expression and self-fulfillment. Possibly they are somewhat more likely to divorce their spouses and less likely to murder them than their unanalyzed counterparts. It would be almost impossible to gather the relevant data and necessary controls to clear up these doubts.

I merely wish to raise the possibility that in at least some instances, a person may be less likely to act out a clearly immoral and socially undesirable wish if he continues to repress it and more likely to lose control if he·becomes aware that he has such a wish. Self-knowledge is not always an unalloyed benefit. Some psychotherapists acknowledge as much in trying to help some patients overcome distressing thoughts by means of what is frankly called *suppressive psychotherapy.*

In any event, since Freud brought the concept of unconscious desires so insistently into general consideration, ethics will hereafter have to wrestle with the complicating assumption that a person may be much more fully and easily aware of some of his wishes and intentions than others; therefore he may be free in choosing certain courses of action but inwardly compelled to undertake others. Hence, there is the possibility that we may have *degrees* of responsibility for our behavior.

MORAL DEVELOPMENT

I can no longer delay talking about another significant development in contemporary ethical thought, which psychoanalysis indirectly helped to come into being: the developmental approach to morality.

Freud did not invent the developmental outlook, but he was one of its most influential advocates and he translated it into vivid terms in his scheme of psychosexual developmental stages. Once one is taken with the power of the idea that the infant grows both steadily in some respects and phasically in others, the conception of an invariant sequence of qualitatively differing steps on a ladder becomes a widely useful thought model. The sequence, oral–anal–phallic–genital is just such a conception of saltatory and emergent development; here periods of rapid change alternate with plateaus, on each of which the child is preoccupied with qualitatively new themes.

While Freud was crystallizing and publishing these ideas about psychosexual development, a group of writers were converging on a structurally similar developmental analysis of morality. In a single decade, Baldwin, Hobhouse, Westermarck, McDougall, and Dewey and Tufts all put forward phasic theories of moral development.[19] Despite their striking convergence, the new ideas penetrated hardly at all into psy-

---

[19]Cf. J. M. Baldwin, *Social and Ethical Interpretations in Mental Development* (New York: Macmillan, 1897); L. T. Hobhouse, *Morals in Evolution* (New York: Holt, Rinehart & Winston, 1906); Edward Westermarck, *The Origin and Development of the Moral Ideas*, 2 vols. (London: Macmillan, 1906–1908); William McDougall, *An Introduction to Social Psychology* (London: Methuen, 1908); and J. Dewey and J. Tufts, *Ethics* (New York: Holt, 1908).

chology or psychoanalysis; I cannot comment on how important an impact they had on philosophy. After Jean Piaget published his own researches on the development of moral judgment in the child, however, the conception started to take hold among psychologists. Piaget admits to having been much influenced by Freud's conception of development, but he also cites the work of Dewey, Baldwin, and Westermarck. The past couple of decades have seen an accelerated interest in moral development, largely sparked by Kohlberg, who in turn was mostly stimulated by Piaget but also by Freud.[20]

All these authors agree in distinguishing three major sequential types of moral thinking. First comes a preconventional, egocentric way of dealing with moral issues, which Piaget called anomic (lawless). There then follows a stage of conventional morality, when the person learns to follow externally prescribed rules of conduct. The third level—which Piaget called autonomous, Kohlberg postconventional, and Dewey and Tufts the level of conscience—characterizes people who have gone through the second stage and have made a personal commitment to an examined set of values. Although Freud never discussed the issues in this way, it is evident that he pictured the young, impulse-ridden child in ways corresponding to the first level. The second level is that of the usual law-abiding but unanalyzed citizen; while the third level corresponds fairly well to the analyzed person who has learned to question rules rather than blindly follow them and to reach rational resolutions of moral conflicts. It also corresponds to what some psychoanalytic writers have called a state of *superego integration,* in which the standards making up the moral agency of the personality are reconciled with the ideals of the conscious ego or self.[21] Kohlberg has by now differen-

---

[20] Cf. Jean Piaget, *The Moral Judgment of the Child* (New York: Free Press, 1932); L. Kohlberg, "Stage and Sequence: The Cognitive-Developmental Approach to Socialization," in *Essays on Moral Development* (New York: Harper & Row, 1981).

[21] H. Hartmann, *Essays on Ego Psychology* (New York: International Universities Press, 1964).

tiated these levels into six specific stages of moral development, plus a final, idealized state.

From this developmental perspective, it is possible to resolve some otherwise perplexing paradoxes. A conventional
morality is a great advance over preconventional, essentially
unsocialized states in which people's conduct is guided primarily by their seeking direct gratification and avoiding punishment. As adolescents begin to grow beyond simple rule
following, however, they usually become rebellious and superficially seem to have regressed to a preconventional status
until it becomes evident that they are learning to listen to a
higher law than custom, guiding their conduct by abstract
principles and ideals such as justice, equality, and truth.[22] To
the average American citizen, thought to be at the "law and
order" stage (the higher of Kohlberg's two kinds of conventional moral reasoning), civil disobedience of a bad law in the
service of an abstract ideal is just lawbreaking, and a principled
protester is not to be differently treated from a common criminal. To see that Mahatma Gandhi and Martin Luther King
were profoundly moral men requires of the evaluator that he
should have developed to a postconventional stage.

Freud's ideal of the well-analyzed person cannot quite be
assimilated to the general conception of postconventional moral
development; for one thing, it omits one major issue that strikes
me as extremely important. That issue is perhaps in the realm
of normative ethics rather than a more structural metaethics.
I am referring to a theme often noted in empirical studies of
moral development: in maturing, a person tends to become
not only more *autonomous* but also more *allocentric* (as opposed to egocentric). I have used the terminology of Douglas
Heath for a distinction that was made long ago by Andras
Angyal in describing two fundamental motives, a trend toward
greater autonomy and a trend toward greater homonomy.[23]

---

[22] It is quite possible that much of the complaint about modern immorality
comes from conventional persons alarmed by these signs of moral development.
[23] Douglas Heath, *Maturity and Competence* (New York: Gardner, 1977);
Andras Angyal, *Foundations for a Science of Personality* (New York: Com-
monwealth Fund, 1941).

David Bakan calls it the distinction between agency and communion.[24] But Freud made no such distinction. In several subtle ways, his theories steadily push us toward the relatively egocentric ideals of personal freedom and self-actualization.

One such subtle influence is the focus of moral conflict on sexuality. Until he put forward the dual instinct theory rather late in his career, Freud constantly portrayed libidinal impulses in the role of antagonist to moral scruples and inhibitions.[25] And even after he formally made the death instinct an equal partner to eros, the cast of his own thinking never quite made the change to conceiving of destructiveness and greed as the major forces needing social control. In 1923 he wrote:

Towards the two classes of instincts the ego's attitude is not impartial. Through its work of identification and sublimation it gives the death instincts in the id assistance in gaining control over the libido, but in so doing it runs the risk of becoming the object of the death instincts and of . . . perishing. . . . It would be possible to picture the id as under the domination of the mute but powerful death instincts, which desire to be at peace and (prompted by the pleasure principle) to put Eros, the mischief-maker, to rest; but perhaps that might be to undervalue the part played by Eros.[26]

All forms of aggression and hatred were transformations, for Freud, of the originally self-directed death instinct. So even destructiveness becomes basically an internal danger.

In another subtle way, Freud's terminology stresses the fact that he thought about people as individuals, a set of patients or potential patients. As I have said, his theory had no place for a social or cultural reality, no way to conceptualize these levels of understanding; its gaze was resolutely intrapsychic. Thus, relations among persons come out as *object relations:* only the patient of concern is a subject, everyone else merely an object in his personal world. Psychoanalysis

---

[24] D. Bakan, *The Duality of Human Existence* (Chicago: Rand McNally, 1966).
[25] *Beyond the Pleasure Principle* (1920), *Standard Edition,* 18.
[26] *The Ego and the Id, Standard Edition,* 19:56, 59.

lacks a social vision in which all members of a society are equally important and valuable participants who owe their (apparent and partial) independence to their interdependent mutual support and to the social fabric of law, custom, and culture. Freud had only a hazy notion of what a metaphor like "social fabric" might mean; he would have suspected that it was only another guise of the protean instinct eros, which binds all things together. However much partial truth this insight contains (and it is the principal message of his social psychology), it dangerously misses the point; for the pursuit of libidinal aims as easily drives people apart as together and can destroy their institutions.[27]

As Roazen[28] has shown, Freud was an old-fashioned liberal in his political, social, and economic thought. That much would be expected from the life circumstances in which he grew up, the bright and upwardly mobile son of an unsuccessful Jewish businessman who had literally come out of the ghetto in his own youth. Further, he lived in a Viennese milieu where all the people he, as a young man, identified himself with were liberals and opposed to the power of the monarchic state. Liberalism was the heritage of the only recently successful bourgeois revolution against the feudal system, codified by such champions of the emerging capitalism as Adam Smith.

As Michael Walzer perceptively remarks,

> liberalism is above all a doctrine of liberation. It sets individuals loose. . . . It abolishes all sorts of controls and agencies of control. . . . It creates free men and women, tied together only by their contracts (i.e., rational agreements). . . . It generates a radical individualism and then a radical competition among self-seeking individuals.[29]

Psychoanalysis is a true child of liberalism and the romantic revolution that followed the Enlightenment. It too marks

---

[27] *Group Psychology and the Analysis of the Ego, Standard Edition*, 18.
[28] Paul Roazen, *Freud: Political and Social Thought* (New York: Knopf, 1968).
[29] M. Walzer, "Nervous Liberals" (Review of *The Neoconservatives* by Peter Steinfels) *New York Review of Books* 26(15):(1979), p. 6.

individual freedom from irrational and unjust restraints as one of its highest values; as a therapy, it tends toward an extreme of individualism. Freud truly believed that civilization would profit from the loosening of neurotic inhibitions, including the conventional overvaluation of authority, and from the freeing of everyone to pursue happiness in his or her own way. Psychoanalysis, again, is only one contributor to a cultural trend of this sort, which Lasch[30] calls the culture of narcissism; but I believe that it shares with liberalism the shortsightedness of giving overriding importance to personal freedom.

Perhaps it is necessary or at least useful to a growing industrial society surrounded by unexploited natural resources to have an ideology and value system that glorify the unrestricted individual, acting to maximize his own monetary profits. Why, then, did not the nineteenth century rather quickly collapse into an anarchic struggle of each against all, given the rapid pace at which the economic power of successful entrepreneurs rivaled the controlling power of the state? The question betrays a subtle carry-over of individualistic bias and a failure to recognize that a society, even one that glorifies a laissez-faire, is to a considerable degree a *self*-regulating system, in which cybernetic control need not take the form of laws and bureaucratically enforced rules. One form of social control is individual morality, which may have been more effective in the nineteenth century than today. The child-rearing practices of the era instilled rather strong moralistic self-controls into the rising class, and such traditional practices change slowly, tending to perpetuate themselves. But as money has become more and more the dominant value and self-indulgent consumption of goods the encouraged proximate goal of life, parents have put ever less emphasis on instilling the old virtues of hard work, thrift, honesty, and decency (not to mention justice and truth). To advocate such values sound quaintly old-fashioned if not positively reactionary in the cynical atmosphere of today.

[30] C. Lasch, *The Culture of Narcissism* (New York: Norton, 1979).

In two of his books, Erich Fromm has stressed the importance of religious orientation, which he conceptualizes as defined by two antithetical ideal types: authoritarian religion, stressing submission to an omnipotent God, with obedience being the primary virtue; and humanistic religion, with self-realization as the primary virtue.[31] I believe that he is correct in seeing these as the antithetical choices available within the psychoanalytic perspective, choices not only of religious orientations but of value systems. The only allocentric alternative to an individualistic emphasis seems to be the traditional submission to secular or divine authority; hence, the moral trend of psychoanalysis subtly encourages an egocentric outlook. In doing so, of course, it harmonizes well with the dominant trend in American culture, which no doubt had a good deal to do with its ready acceptance here. In their hopes for a more benign and enlightened substitute for authoritarian religion, both Fromm and Freud have overlooked one important and troubling finding of cognitive-developmental psychology. Kohlberg alleges a rather close relation between cognitive and moral development, such that attaining Piaget's highest stage of cognitive growth is a necessary but not sufficient condition for advancing to postconventional, principled morality. In either Fromm's conception of a humanistic religion or Freud's rational, reasonable morality of the well-analyzed person, postconventional moral thinking requires a degree of intellectual attainment that only a tiny percentage of human beings possess.

Here is the joker in the seemingly optimistic deck offered by developmental psychology. Everyone starts at the bottom of the ladder and must (we are told) go up the same steps in the same order. Some people progress faster than others, who may become fixated at a lower level than necessary, but therapy and more enlightened education can help many of the laggard to keep moving—up to limits set by personal potentialities. That last qualification does not sound so bad until we

[31] *Psychoanalysis and Religion* (New Haven: Yale University Press, 1950); *You Shall Be as Gods* (New York: Holt, Rinehart & Winston, 1966).

learn how few people have been found at advanced levels of moral development. True, only impressionistic data are available; Kohlberg's test of moral reasoning has not been given to a scientific sampling of any general population. Yankelovich and I did, however, give Loevinger's[32] Sentence Completion Test for assessing her closely related conception of ego development (which has also been found to be moderately well correlated with Kohlberg's test) to a national probability sample of American youths between the ages of 16 and 25. We found only 15 percent of our group to be at Loevinger's "conscientious" level, where (she says) behavior is controlled by well-internalized inner standards. Loevinger distinguishes three higher levels which sound even more clearly like Kohlberg's postconventional level, but we were able to find fewer than 5 percent of our sample who scored at these top levels.[33]

Even granting certain methodological weaknesses of our survey and the only partial equivalence of moral and ego development, all available evidence suggests that a small minority indeed of contemporary Americans have attained postconventional morality and that the intellectual prerequisites may be responsible. One cannot safely plan for a new, good society on the assumption that liberated but principled people with egos well integrated to superegos are going to prevail; according to this argument, the effort to replace simplistic, inhibitory morality based on religion would not work, for lack of a sufficient intellectual base, even if the benefits of psychoanalysis could be very widely spread.

The gloomy picture in the preceding paragraph brightens appreciably once several assumptions are made explicit. First, not even Kohlberg has claimed that his method provides an assessment of more than moral *reasoning;* though it is easy to equate thinking or talking about what is right with behav-

---

[32] J. Loevinger, *Ego Development: Conceptions and Theories* (San Francisco: Jossey-Bass, 1976).
[33] R. R. Holt, "Loevinger's Measure of Ego Development: Reliability and National Norms for Short Male and Female Forms," *Journal of Personality and Social Psychology*, 1980,39,909–920.

ioral probity and integrity, we have no logical or empirical warrant to do so. Second, there exist no normative data to back up the claim that a large proportion of humankind is genetically limited in intellectual potentialities to the indicated degree. It has *not* been demonstrated that the usual kinds of IQs are relevant to the capacity to live morally good lives in the world of the future.

Nevertheless, I believe that we are faced with a dilemma. The crisis of a world poised on the verge of thermonuclear suicide is undeniably a moral as well as political one, which cannot be solved by the frequently prescribed return to religion. Even personally devout Christians like Presidents Carter and Reagan have not felt morally restrained from threatening to go to war in defense of rather limited "national interests" in an age when any conflict involving the great powers can easily turn into an atomic holocaust leading to the extinction of humankind. It does not seem that an effective morality adequate to the needs of today and of the future can be grounded in Christianity or any other extant world religion; if anyone is founding a new religion—or any other social institution for the control of human behavior—with the emotional power and sufficiently widespread appeal to serve as the vehicle of the needed morality, I have yet to hear of it. Henry A. Murray is among the very few who have even uttered a clear call for such a social invention.[34] It would have to accomplish the basic purpose of bringing almost everyone up to the level of rule-following morality, providing an improved set of rules more conducive to the survival of life, while still allowing as many of us as can to continue to develop beyond rule following without having to reject the religion. That is a staggering intellectual challenge!

If I may for the moment slip into a prescriptive role, I think the morality of a sustainable future must be lovingly

---

[34] H. A. Murray, "Beyond Yesterday's Idealisms" in *Endeavors in Psychology: Selections from the Personology of Henry A. Murray,* ed. E. S. Shneidman (New York: Harper & Row, 1981).

allocentric, nonsubmissive, and nonauthoritarian. It will be a morality of interdependence, of mutual need and mutual aid, and one of responsibility based on an ideal of universal empathy—a capacity to identify oneself with all living species present and future. The debate on nuclear power is forcibly bringing to our awareness the fact that our contemporary decisions will have profound repercussions on the quality of life for countless unborn generations. And the endless interrelatedness of the crises and catastrophes that threaten us implies the need for an ecological ethics of implication. That is, it is not enough to base conduct on abstract moral principles, for even such a lofty principle as the sanctity of human life can lead to the excesses and dangers of the "right to life" movement. We need a systemic morality that recognizes that actions have no absolute moral value isolated from their systemic context. The moral worth of an act, no matter how well intentioned and well-based on an abstract moral principle, cannot be judged outside of its situation *or* of its total context in human history— its remote as well as its immediate consequences. If that seems to perfectionistic a standard, I can only reply that I thought it was precisely the function of moral ideals to portray aspects of perfection. But how to do so without the destructive consequences of what we call perfectionism—there's the rub.[34a]

FREUD AND THE EMERGENCE OF A NEW WORLD HYPOTHESIS

Unhappy though I am to leave such important issues in so unresolved a state, I can do no more with them. I want to attempt a more hopeful conclusion by drawing attention to one final indirect effect of psychoanalysis upon the philosophical—

[34a] According to Carol Gilligan, *In Another Voice* (Cambridge: Harvard University Press, 1982), women generally think about moral issues contextually and in terms of care and responsibility, while men tend to stress abstract justice. If her data can be generalized, the needed changes described above could be attained without the widespread structural growth that seems so hard to achieve, by society's giving the alternative moral voice of community a more prominent role in the moral consensus. She stresses that it is by no means exclusive to women. (Note added in proof.)

including the ethical—thought of our time. That can be expressed as Freud's influence on the emergence of a new world view, or world hypothesis.

World Hypotheses

After hesitating a good deal over the choice of a term, I have settled on *world hypothesis* in place of the more easily accepted and widely used term *world view*. Precisely those seductive features have made the latter phrase spring into vogue during the past few years, with a consequent diffusion and dilution of its meaning. Meanwhile, a modest renewal of interest in the work of Stephen C. Pepper (1891–1972) is currently taking place, though his concept of world hypothesis does not yet have any wide currency.[35] It is nevertheless just what I was looking for—a conception of basic philosophical assumptions that underlie, support, and subtly guide many aspects of our conscious thoughts and behavior, though we may not be focally aware of having adopted them. Whether a person has ever given philosophy as much as a passing glance, he cannot help picking up and incorporating into his own outlook the prevailing definitions of reality (ontology), ideas about the place of man in the universe (cosmology), and conceptions of knowledge (epistemology). In short, we are all metaphysicians, implicitly if not explicitly, and we must take some stand on ethical principles as well.

Pepper's peculiar merit, in contrast to others who have surveyed and classified types of metaphysical positions, is that he searches for and finds a small number of independent, more or less equally comprehensive and adequate philosophical positions, each stemming from a distinctive *root metaphor.*

A man desiring to understand the world looks about for a clue to its comprehension. He pitches upon some area of commonsense fact and tries if he cannot understand other areas in terms of this one. This original area becomes then his basic analogy or root metaphor.[36]

[35] S. C. Pepper, *World Hypotheses: A Study in Evidence* (Berkeley, Calif.: University of California Press, 1942).
[36] Ibid., p. 91.

From this root grow categories, used as "basic concepts of explanation and description" and applied to all other areas of fact and speculation. The most fertile root metaphors generate the most adequate explanatory systems, or world hypotheses. With remarkable neutrality, he examines the range of metaphysical positions and settles upon four as "adequate"—formism (e.g., Plato), mechanism (e.g., Locke), contextualism (e.g., Dewey's pragmatism), and organicism (e.g., Hegel's idealism). He himself later added a fifth, which he called selectivism. Though I have cited approximate examples, none of these world hypotheses is exactly identical with the philosophical system of any one man. Instead, it is an abstracted, internally coherent, implied position, a kind of ideal type toward which the "strain toward consistency" in human thought impels us.[37] At the same time, any given philosophical work is the outcome of various other influences, so that it is likely to be a compromise formation (in Freud's phrase) to some degree, even to the point of an attempt to hybridize fundamentally incompatible systems of ideas. Cartesian dualism, for example, is such a conjoining of two worlds—one of matter, one of spirit—each with its world hypothesis: mechanism and animism.

I hope to show (though more incidentally than through exposition) that, despite their somewhat shadowy and conceptual existence, world hypotheses have an important kind of reality in intellectual history.

FROM ANIMISM TO MECHANISM

The first major world hypothesis, which dominated European thought for many centuries, does not appear in Pepper's list of four, since he did not consider it adequate. He calls it *animism*, though in the form that is most familiar to us it is theology or religious supernaturalism. As Freud noted, it is the oldest theory of the world; still vigorous in his time (and

---

[37] The coherence of a world hypothesis is not that of a formal system and should not be taken to mean the rigid lucidity of deductive interconnectedness.

ours), it was the target of most of his polemical writing of a philosophical character.[38] Its perennial appeal despite its philosophical weakness, Pepper believes, arises from its being based on the most appealing of root metaphors: the human being. The spirits primitive people imagine in an effort to comprehend the events of nature are transparently projections of themselves, and no matter how profound, elevated, or sophisticated the theology, the concept of God can be traced back in an unbroken lineage to simple projective animism. That does not, of course, invalidate it. One secret of the staying power of this world hypothesis is its imperviousness to evidence. The principal inadequacy with which Pepper charges it, however, is its lack of precision. Since it begins by postulating a supernatural world beyond human ken except by revelation to persons who are able to persuade others that they have privileged access, it tends to be authoritarian and hostile to skeptics. The only limits on what can be asserted are set by the cognitive consciences and self-discipline of religious leaders and spokesmen.

Parenthetically, I should note that the first adequate world hypothesis, which Pepper calls *formism*, was developed by Plato and has long enjoyed the respect of many philosophers and educated people generally. Yet it was never able to command the adherence of a large enough segment of the world of ideas (except perhaps in the classic era) for us to be able to call it the dominant theory.

Let me briefly remind you of the major tenets of the animistic world hypothesis. Cosmologically, the ostensible center of everything is God, who created and oversees it, intervening at will. But it is also an anthropocentric universe: God made the world for man and set our habitation, the earth, in its literal center. However omnipotent and omniscient God may be, he is remarkably attentive to and solicitous of every mortal one of us. What is most real is the spiritual realm—our immortal souls, God, and the afterworld. The world of this life is a passing

---

[38] *Totem and Taboo* (1912–1913), *Standard Edition*, 13.

show, of purely temporary importance, a brief prelude to an eternity in heaven or hell. Its only importance is that our behavior in this world may determine where we spend the infinite future. Man's knowledge is limited, but God is omniscient; the most important things we know by faith, not by observation or reasoning. If you would seek knowledge, go to the wise men or the sacred books, the traditionalistic sources of truth; it is impious folly to try to discover truth on your own. Religion can teach us the meaning of life: we are here for a purpose, because God put us here for ends of his own, and everything that happens is part of His (usually inscrutable) grand plan in a teleologically ordered universe. He has laid down moral rules for the guidance of our conduct, also; we must obey his commandments. Since we have immortal souls, not subject to the laws that govern brute matter, we are free to choose which path to follow and so are morally responsible for our acts.

As E. A. Burtt has brilliantly shown, the new physics of the seventeenth century brought with it a revolution in ways of conceiving of the world.[39] The new formative image, the mechanical clock, was an important invention of the preceding medieval centuries. Machines were the main cultural artifacts usefully supplied by the new science, and the spring-driven clockwork made possible the construction of clever automata, mimicking the movements of living beings. The machine remains a principle guiding figure of our culture; witness the readiness with which it is invoked figuratively when we refer to the structure of anything from the galaxy to the atom, and particularly the human being and human institutions.

Even though such pioneers as Kepler and Newton were devoutly religious, they created a picture of the universe as a majestic clockwork requiring nothing more of God than that he create the machinery and serve as its prime mover. By the late-eighteenth-century period of the Enlightenment, it became increasingly feasible to encompass all explanatory

---

[39] E. A. Burtt, *The Metaphysical Foundations of Modern Physical Science*, 2nd ed. (London: Routledge & Kegan Paul, 1932).

knowledge about the universe and man without needing to invoke God at all. He was read out of existence by such thoroughgoing materialists as Baron d'Holbach and (in Freud's time) Büchner, who were not afraid to be outright atheists. A complete and consistent world hypothesis of reductionistic mechanistic materialism emerged. It did not need to be adopted by scientists generally to have a profound influence upon them, however piously they might attend religious services and devoutly believe in God. It was no longer necessary to assume a spiritual reality behind the material (though one might if one liked). Simple location in time and place became the hallmark of the real, and the hard, tiny, grittily enduring atom was its very epitome.

In cosmology, a new grandeur of immensity, provoking a rather dismal sense of wonder, had to serve in place of religious reverence: the universe was far larger, emptier, lonelier, and more indifferent than before, and human beings seemed humble and trivial indeed, living on a minor planet of a modest, ordinary star without an easily specifiable place among billions of neighbors in a galaxy undistinguished among countless others. Our lives are the briefest flicker in a stony, insensate eternity of time, devoid of any special significance or claim to notice. It became meaningless to ask for the meaning of life; purpose had no more place in this austere and flavorless world, governed by abstract, quantitative laws, than did spiritual entities like souls, or secondary qualities like colors and feelings.

Alfred North Whitehead has shown how the new scientific world hypothesis required Locke's and Descartes's doctrine of primary and secondary qualities.[40] If what was essential to science about an object was its quantifiable mass and its location in space and time, then by a procrustean logic all its other properties (color, sound, odor, etc.) were mere secondary qualities resulting from the operations of minds and not of interest to science. That is, the grand objective order of nature

---

[40] A. N. Whitehead, *Science in the Modern World,* rev. ed. (New York: Mentor, 1952).

had been formulated without them, in terms of only the primary qualities, which soon were known merely as quantities.
Such a drab and abstract nature would hardly interest anyone
but a scientist, yet it is wonderfully useful to him. Precisely
this impoverishment of their subject matter freed scientists to
work with growing success, providing humankind with many
concrete benefits.

Note how central was the role of analysis in this system.
Newton succeeded so well because of his analytic ability to
extract from apples and luminous heavenly spheres alike their
elementary properties of mass and movement in space and
time. Generations of scientists since have followed his methods, for Newton seemed to be saying: to succeed, simplify;
assume an isolated system; find the basic elements; experiment to find their causal interrelations; seek the mathematical
formula that will explain how they operate conjointly; and you
have the answer—scientific law. The important new method
of experimental control means taking pains to narrow one's
field of vision to see more surely by looking at less; the subject
matter is isolated from its natural setting by careful artifice.
The thematic emphasis of atomism was highly congenial to
the analytic scientific mind, though logically separate.

Despite what Freud was to call the cosmological blow to
man's narcissism delivered by Copernicus in destroying the
geocentric world picture, science gave reason for pride and
hope; for human knowledge, despite its fallibility, now becomes in principle indefinitely extensible if we only hew to
the rules of the scientific method. We must receive all traditional wisdom with skepticism and demand to be shown evidence—that is, controlled observations. These are then organized by clear logic into testable laws. It was consistent with
this world hypothesis to rely on the cool, lucid intellect to solve
moral as well as empirical problems. But since all events are
rigidly determined down to their tiniest details by gapless
chains of cause and effect, freedom of choice is an illusion.
By the late nineteenth century, morality had become a subject
for dispassionate scientific examination, for no department of

life or experience was exempt. Indeed, many tried to produce a new, scientifically based ethics, though without success.

Lewis Mumford has brilliantly argued the case that there has always been a close connection between materialistic, mechanistic science and political authoritarianism. It is not just in the Reagan budget that by far the greatest expenditures for research and development are allocated to the war machine; the latter has always been science's most faithful patron. The hierarchical, machinelike structure of bureaucratic institutions requires of its human components that they perform as much like interchangeable mechanical parts as possible. The ultimate extension of this dreadful but powerful fantasy is what Mumford calls the megamachine—the entire society in a totalitarian organization, rolling along over everything in its path with one dictatorial despot in the driver's seat: Hitler conducting the Holocaust through his order-taking intermediaries.

Clearly, I do not mean to imply that everyone who is true to the ideals of nineteenth-century reductionistic science will end up doing the grisly "experiments" of the Nazi doctors. Human beings are not wholly consistent, and very few people have clearly espoused the entire mechanistic ethos. Yet the fact that all the elements of this ideological configuration exist and fit naturally together creates a subtle, implicit pressure to believe in most aspects of the whole system, or to act as if you did.

FREUD, RELATIVISM, AND PRAGMATISM

As soon as the mechanistic world picture began to unfold, it met opposition. The main new philosophical alternatives for a couple of centuries were idealisms, equally extreme attempts to reduce the material to the mental or spiritual instead of the reverse. In the hands of Hegel, a third and equally adequate world hypothesis took shape, which Pepper calls *organicism*.[41]

---

[41] Like formism and contextualism, it never captured the popular imagination as animism and mechanism have done.

One can look on the romantic movement as another reaction against the austerity of the ethos of mechanistic science, and it is important to remember that most scientists managed as human beings because they were brought up in a culture where religious indoctrination began very early and the school system was pervaded by humanism.

It is obvious that in many ways, Freudian psychoanalysis was a natural outgrowth and expression of the mechanistic world hypothesis, especially so in its metapsychology. To be sure, Freud had approximately as deep commitment to a contrasting set of ideas from the softer tradition of humanism, the central core of ideas and values conveyed by the humanities. His theories were always more or less successful struggles to synthesize the themes and outlook of humanism with those of mechanistic metaphysics. Humanism is not a consistent Weltanschauung but a valuable partial set of themes opposed to and threatened by both the supernatural and the mechanistic world views. Any ideology broad and loose enough to attract the fealty of thinkers as diametrically opposed as Carl Rogers and B. F. Skinner can hardly be integrated or a workably consistent set of philosophical footings.

Fairly early in the nineteenth century, the mechanistic world hypothesis began to be challenged and stretched by some developments within the hard sciences themselves. In a growing variety of disciplines, notably crystallography and morphology, some concepts of structure and form began to be necessary. Ecology, in some ways the most typical scientific embodiment of an emerging emphasis on contexts and whole systems, had its beginnings in some of Humboldt's shrewd observations about landscapes as habitats, and in much that Darwin wrote, for example, about the evolutionary fitting of species to their environments. A number of facts began piling up in physics that were embarrassing to the Newtonian cosmology. All along, of course, advocates of the religious–supernatural world view continued to thunder against scientific materialism, and we should keep in mind the fact that the great majority of humankind has always held a basically dualistic, supernatural set of metaphysical assumptions—to this very day, as the challenge of creationism attests.

In many respects, the intellectual crisis of the nineteenth century that was precipitated by a reluctant Darwin only continued a task that had been begun by the Enlightenment and the romantic era, that of undermining settled, traditional, absolutist ways of thinking. As the intellectual bankruptcy of dogmatic doctrines became evident, one or another form of relativism took its place in the minds of literate people. A student of intellectual development in contemporary young people, W. G. Perry[42] has regularly found the following sequence. Naive youngsters, so unquestioning of their basic beliefs that they may be called dogmatists, in college first encounter multiplicity: a diversity of people, ideas, systems of values, ways of thinking, and so on. After several intermediate positions of attempting to come to terms with multiplicity, the student typically adopts the relativistic stance that "everyone is entitled to his own opinion."

Without too much oversimplification, the history of Western culture can be said to have followed a similar course during the past four centuries. In the era of great explorations, travelers brought back marvelous tales about the multiplicity of human beings and their cultures; eventually the pursuit of such exotic curiosities became anthropology. The particularistic romantic temper was especially fascinated by the diversity in the ethical systems by which the world's peoples have lived, and many people were at first shocked to discover that ideals which they had always taken as universal were ethnocentrically limited. If there were as many moral perspectives as cultures, the disquieting thought arose, perhaps all were equally valid. Thus cultural relativism in ethics was born.

But moral relativism also raised the possibility of what Piaget calls decentering (see below). That is, once we realize that there are various other cultures besides our own, with

[42] W. G. Perry, *Forms of Intellectual and Ethical Development in the College Years* (New York: Holt, Rinehart & Winston, 1970).

somewhat different moral views, we become aware of the possibility that what seems absolutely right and wrong to us may be so only because of our cultural conditioning. (We do not have to accept the premises that different cultures have fundamentally different moral views, or that our moral views would have been different if we had had different accidents of birth and rearing. The possibility alone is upsetting to orthodoxy and absolutism.) We can now envisage a new stance, previously invisible to us, outside of and perhaps above our culture, suspended in some hyperspace in which cultures are arrayed. From this new imaginary standpoint, it is possible to raise such questions as: Are any of the moral systems or principles of any cultures absolutely right or wrong? Could another sentient and intelligent creature have quite a different moral sense or indeed none at all? Granted, there are no easy answers to these and similar questions, as D. H. Munro[43] argues. Nevertheless, it was a striking intellectual achievement to have attained the decentered position, which alone permits questions of this type to be asked.

Consider also the relation between pragmatism and developments in mathematics. With the development of non-Euclidean geometries early in the nineteenth century, it became evident that a geometry was one of a number of possible formal systems, none of which had the absolute and eternal verity that had long been claimed for Euclid's. Likewise, language and mathematics generally are formal systems of this kind; their value is contingent upon the human purposes to which they were harnessed. C. S. Peirce could see that and put forward the first pragmatic theory of truth precisely because of his attainments in symbolic logic.[44]

This common theme of liberation from absolute, fixed dogmas swept through many fields of thought during the nineteenth century. Before the Darwinian revolution, the general

[43] D. H. Munro, "Relativism in Ethics," in *Dictionary of the History of Ideas*, ed. P. P. Wiener, vol. 4 (New York: Scribner's, 1973).

[44] P. P. Wiener, "Pragmatism," in *Dictionary of the History of Ideas*, vol. 3.

conception of the biological universe was a fixed and stable one: the forms of life we know were created at one time and would persist unchanged until judgment day. Darwin not only upset the authoritarian dictates of the Christian churches about the "kingdom" of living things but also substituted a view of ceaseless change with no logical goal or endpoint.

Freud did his best to apply this Darwinian view to human psychology. He did not conceive of a fixed and eternal human nature but saw man as a continually evolving creature with different instincts and potentialities than his remote ancestors had had. In that respect, he was not very different from many of his contemporaries, for all ideational shores were wet by the rising Darwinian tide.

Freud did, however, make a more distinctive and personal contribution to the growth of relativism and the new possibilities of decentering, for what he called the three "blows to man's narcissism" delivered by Copernicus, Darwin, and himself may also be pictured as three stages of decentering, each freeing us from an old dogmatism and introducing a type of relativism. When Copernicus dethroned the earth from its place at the center of the physical universe, he gave us a vaster vision: long before space travel became a literal reality, people were freed to roam in imagination through a world of many worlds, a universe of universes, where before we had been confined within a single small and parochial one. When Darwin showed us that we were a species like any other, with a common ancestry and a future of continued modification, he greatly enlarged our self-knowledge, freeing and inspiring all the biological sciences. Our increased capacity to free ourselves from disease is only the most obvious benefit.

Freud likewise brought humankind unpalatable but nourishing truth. He correctly saw that many people would feel threatened by being told that they were not entirely masters of themselves, that they could not easily know even their own minds because of the veils of repression, and that most of us spend much of our lives endlessly enacting versions of simple but unconscious dramas from our early childhoods. In works such as *Totem and Taboo* he showed the imprint of impulse and defense, operating without the conscious will of the actor, on systems of religious and philosophical ideas as well as on neurotic symptoms. Here, then, was a psychological relativism growing up alongside the economic and sociological relativism of his contemporaries. The concept of rationalization—that we hold certain ideas not for their superficial and ostensible content but because they satisfy the obscure imperatives of desire and fear—had a swift and enormous vogue if an unsettling effect. Who could be sure that one's beliefs were not clever fronts for wishful conclusions? One could not assume that everyone else shared the same psychic reality; what seems real for me is to you transparently determined by my unconscious wishes and defenses. The unremembered traces of infantile traumas and fixations leave a different coloring on everyone's personal spectacles.

In all the forms of relativism I have briefly sketched, the phenomenon of what Hofstadter[45] calls the "strange loop" occurs: the relativist eventually discovers that he is sawing off the limb on which he himself is sitting. If each culture seems to have its own moral system, which anthropologists tell us fits it and seems necessary to it, perhaps all are equally valid or invalid—including the system of values within which I am making the present value judgment! If all ideological systems are ultimately determined by the economic system of producing and distributing wealth within which they grow up, then how could Marx himself claim to stand outside? Why are his own ideas—indeed, the very doctrine of economic determinism—not subject to the same caveat? If conscious ideas and arguments are often rationalizations for decisions made on the basis of unconscious wishes or fears, what guarantee is there that Freud's own theory—including the concept of rationalization itself—was not a vast rationalization? The problem of relativism, therefore, is that of the strange loop of self-refer-

[45] D. R. Hofstadter, *Gödel, Escher, Bach: An Eternal Golden Braid* (New York: Basic Books, 1979).

entiality, which produces so many of the familiar logical paradoxes. The oldest is that of Epimenides the Cretan, who baffled his listeners by declaring, "All Cretans are liars." Or, more baldly put, "This sentence is false." As a moment's reflection shows, it can be neither false nor true!

Relativism is hardly a comfortable position; it is inherently unstable. Perry finds that some college students lapse from it back into the values and outlooks with which they first entered the new milieu, while others move ahead to what he calls "commitment," the affirmation of an examined and personally integrated set of basic principles and goals. Nevertheless, near the end of the nineteenth century, a group of philosophers were able to transform relativism into a vigorous philosophical school, pragmatism, that ultimately constituted a new world hypothesis as adequate as the three that had preceded it.

Pragmatism, according to P. P. Wiener,[46] has five major components: "(1) a *pluralistic* empiricism or method of investigating. . . ; (2) a *temporalistic* view of reality and knowledge as the upshot of an evolving stream of consciousness . . . or of *objects* of consciousness. . . ; (3) a *relativistic* or contextualistic conception of reality and values in which traditional eternal ideas of space, time, causation, axiomatic truth, intrinsic and eternal values are all viewed as relative to varying psychological, social, historical, or logical contexts. . . ; (4) a *probabilistic* view of physical and social hypotheses and laws in opposition to both mechanistic or dialectical determinism and historical necessity or inevitability. . . ; (5) a secular democratic individualism."

Pepper calls the world hypothesis of the pragmatists *contextualism*. The previous summary hints at some of its strengths and hospitality to a new temper that began developing in science around the turn of the twentieth century. If an idea can be fully understood only in some context—for example, as someone's defense against anxiety, as an instrument of oppression, or as a gesture of rebellion against a social norm—any of these relativisms may be taken as a way of sensitizing us

---

[46] Wiener, "Pragmatism," p. 553.

to the ways parts assume special meaning depending on the wholes in which they are embedded. Context thus can take on ontological significance; pattern gains a dignity and centrality it could never have in mechanism.

The very instability of relativism can be converted into a metaphysical principle. The root metaphor of contextualism, says Pepper, is the experienced event, which William James described as part of a stream of consciousness. One needs verbs, not nouns, to discuss the fundaments of reality. This is a philosophy of becoming rather than being, open to continual and basic change:

Change in this radical sense is denied by all other world theories. If such radical change is not a feature of the world, if there are unchangeable structures in nature like the forms of formism or the space–time structure of mechanism, then contextualism is false.[47]

Freud took little interest in pragmatism, showing much less openness to James's ideas than the American did to his. I don't know whether he would have been more amused or indignant if he had been told that part of his contribution to the history of ideas was to prepare the way for this philosophical development, this new world hypothesis to rival mechanism as a foundation for science.

Not that pragmatism (or contextualism) ever seriously threatened to displace mechanism in widespread appeal to scientists. Ironically, though in many respects the world hypothesis of mechanism is thoroughly out of date and far less congruent with modern physics, biology, computer science, and a good many other disciplines than contextualism, its metaphysical assumptions still silently suffuse many aspects of the contemporary ethos. In many universities, methodologists teach a conception of science that stresses analysis to elements as the primary method; reductionism as an ideal; an implicit conception of matter—especially in the form of its subatomic particles—as the ultimate reality; and the classical

[47] Pepper, *World Hypotheses*, p. 234.

experiment as the model of investigative procedures, with values carefully segregated from facts and so far as possible excluded entirely from the realm of science. The computer, far more sophisticated than the old-fashioned clockwork but still a machine, seizes the imagination of our bright young people and bids fair to become the root metaphor of an updated new version of mechanism. Despite all the dazzling achievements of high technology, its underlying philosophical assumptions are rigid, cold, and authoritarian; moral values have no natural roots in them. These assumptions will not do as the foundations of a future world fit for ordinary organisms like human beings.

TOWARD A NEW, INTEGRATIVE WORLD HYPOTHESIS

Contextualistic metaphysics is evidently quite congruent with some versions of modern quantum theory, but not—despite the similarity of terminology—with relativity. Einstein never liked that term, which was introduced by Max Planck for what Einstein himself preferred to call *invariance theory*.[48] His choice of terms indicates his commitment to finding just those "unchangeable structures in nature" that contextualism cannot tolerate. Far from implying merely that "everything is relative," Einstein's work made physics a simpler, more consistent, and universally applicable theory. Einstein did this through the very device of giving up Newton's absolute framework of a fixed reference system of space and time, though intuitively that feels more basic and necessary to many people than preserving a set of invariant equations for the laws of physics. We just have to keep our priorities straight: the most important thing is to know that the basic physical laws will always be the same regardless of the circumstances under which observations are made.

How did Einstein reformulate the laws of nature in a form

---

[48] G. Holton, "Einstein's Task as a Universe Builder: His Early Years," *Bulletin of the American Academy of Arts and Sciences* 33:1(1979):12–28.

that is independent of any particular standpoint, freeing them from relativism? Be developing sets of mathematical transformations that systematically take into account the circumstances under which observations are made and adjust them accordingly. If you are familiar with factor analysis, it may help to think of it this way: represent space and time geometrically as a four-dimensional reference system, and the transformation equations become a way of rotating the axes of this coordinate system so as unfailingly to achieve the same structure of relationships among your data, no matter what form they originally had. It is strictly analogous to the successful effort of a factor analyst to attain a common, simple structure for sets of data based on various batteries of tests and different samples of subjects.

With this example, I want to suggest that Einstein's solution to the problem of relativism is extremely powerful and flexible and applicable to many problems in psychology. Consider, for example, Piaget's work on children's perspective taking. In one of his classical experiments, children who lived around Lac Leman in Switzerland were shown a miniature construction or model of the lake and its surrounding mountains. "If you stand on the north shore of the lake," Piaget would tell a child, "this is how the mountains look; how do they look to a child standing over there, on the eastern shore?" The youngest children would say, in effect, "The same way they look to me, 'cause that's the way they are." When a major step of cognitive development takes place, the child becomes aware that how the mountains look depends on your point of view, and that different people can have different perspectives. Piaget saw, however, beyond the implied relativism to the fact that there is a single real structure to the lake and mountains, which *is* recoverable from the many reports if they are subjected to a set of transformations. Getting out of one's unquestioned position in the center of the universe, or *decentering,* takes a long time, for the process has to be repeated many times in increasingly abstract contexts.

In the experimental psychology of human perception, a similar story has been emerging, notably in the study of the perceptual constancies. Psychologists' curiosity was initially attracted by the observation of certain striking achievements of our senses in constructing a steady world. For example, white paper continues to look white in the moonlight, when it reflects to the eye far less light than the blackest paper does in direct sunlight. Even more astonishing was the realization that our eyes provide us with constantly moving scans of the surrounding world, which nevertheless seems to hold still whether we stand still and stare fixedly or move eyes, head, and body in complex and jerky ways. It turns out that the brain computes a set of transformations of these fluctuating, relativistic inputs in order to emerge with the perceptual invariances that we experience as a stable external world.

Freud, too, went beyond psychological relativism in his discovery of transference and countertransference.[49] Transference is so called because the patient tends to transfer to a contemporary figure, notably the analyst, emotional reactions that properly belonged to another person in his own past.[50] The reaction is mediated by the *symbolic equivalence* of the two persons, an interesting generalization of Freud's original discovery of the way to interpret dreams: they must be subjected to a process of semantic (and figural) transformation, following the rules of primary-process thinking. In a more than trivial sense, I believe that there is a parallel between these transmutations of meanings and forms and Einstein's mathematical transformations; both achieved invariant order out of bewilderingly particular or paradoxical data.[51]

---

[49] E. Erikson, "Psychoanalytic Reflections on Einstein's Centenary," in *Albert Einstein, Historical and Cultural Perspectives,* eds. G. Holton and Y. Elkana (Princeton, N.J.: Princeton University Press, 1982).

[50] It is instructive if saddening to notice how often transference is construed— even by psychoanalysts themselves—as meaning simply falling in love with your analyst. Here Freud's own sexual reductionism joined forces with the regressive temptation to which thought is always vulnerable. That tendency itself, incidentally, is another manifestation of the same thought pattern as transference, which Piaget called "assimilation."

[51] More precisely, those of Lorenz and Minkowski.

Freud went a crucial further step into relativism by recognizing that he himself was subject to the same process of assimilative distortion; he called it countertransference. Here he moved decisively away from the naive absolutism of the assumedly objective expert; the psychoanalyst's perceptions and conceptions of the patient are likewise vulnerable to or are shaped by the unique life experiences that constitute his special perspective. Is that not a relativistic chaos, in which no one knows where the truth lies, if indeed one can now believe in the very possibility of valid human knowledge? It would have been, had Freud not seen a practical way out. The psychoanalyst must himself be subjected to successful analytic therapy. Such treatment does not claim to work by removing all conflicts or by lastingly opening the doors of the unconscious. Instead, the training analysis helps the neophyte learn his own characteristic ways of distorting reality, so that he can take appropriate distance from them and learn compensatory ways of correcting them. That comes close to being a set of transformations that yield an invariant lawfulness and the possibility of approaching true knowledge despite the facts of psychological relativism. It is an imperfect means of promoting decentering, but when it works well it illustrates the principle.

To return to philosophy for a moment, the process I have been describing can be called giving up naive realism for relativism and replacing that by critical realism. It is often difficult for people to grasp the difference between the two kinds of belief in something solidly real and fixed in the universe. The naive form, which I have also called "absolutism," simply accepts things the way they appear to us from whatever standpoint we happen to inherit or find ourselves in and assumes that the resulting reality is absolutely trustworthy, because "seeing is believing." As we get older, we learn to be wary; as we discover that our eyes can deceive us, and that things do look different to different people, we gain a degree of sophistication or decentering. But it amounts to a further achievement in decentering to realize that, outside all the vantage points of all the individual observers, there is a constant and asymptotically approachable framework of reality. We cannot always—perhaps never—see it directly but we can construct it. The qualitative operations of judgment used by the critical-realist philosopher are again analogous to the quantitative operations of the Lorentz transformations that Einstein rediscovered to save the laws of physics.

In an initial enthusiasm for Einstein's principle of finding invariant structures of laws by means of systems of transformations, I thought that this discovery might serve as the germinal cell for yet another world hypothesis. (It does not seem quite close enough to commonsense experience to be spoken of as a root metaphor.) But in reading over Burtt's[52] appreciative critique of *World Hypotheses,* I was struck by an important weakness of the book's argument. Pepper argues that all eclecticism is confusing and that root metaphors must not be combined. That is, any attempt to combine in one system ideas that stem from different root metaphors will fail, he claims, because they will prove fatally incompatible. As Burtt notes, to have proved such a point Pepper would have had to have the metatheoretical advantage that only another world hypothesis could have provided, and Pepper was quick to deny that he had any such ambitions.[53]

The root metaphor hypothesis is indeed convincing, but as a bit of the psychology of ideas, not as a methodological principle. Such metaphors must be discovered by a creative— perhaps artistic—process of intuitive discerning, which has not been reduced to principles that can be reliably used by others. More to the point, no genetic analysis has probative value. We are left, then, with the ancient criterion of consistency. Surely it is desirable to strive for coherence and lack of contradiction. But the world is extraordinarily rich and various; why, then, should we not draw on more than one strand of experience or common sense in building a world hypothesis

[52] E. A. Burtt, "The Status of 'World Hypotheses,'" *Philosophical Review* 52(1943):590–601.

[53] Later, he did attempt to state his own fifth world hypothesis. Pepper, *Concept and Quality: A World Hypothesis* (LaSalle, Ill.: Open Court, 1967).

as long as we do so carefully and with particular concern not to entail subtle forms of contradiction?

The ideals of simplicity and adequacy lead us in opposite directions. In a charming little prose poem Pepper tells us how much stronger, for him, was the emotional pull toward simplicity:

> . . . it occurs to me
> That when we find the truth
> It will be as simple as the blue sky of morning.[54]

This appealing image in a way expresses the esthetic ideal that was so important to Einstein. Yet such a tranquil vision has no greater claim to be a sure guide to the truth than Darwin's tangled bank—itself a strong claimant to be the root metaphor of the ecological outlook.[55]

Pepper erred, I think, in failing to distinguish the feckless eclecticism of an intellectual pack rat from true synthesis, which is as demanding of disciplined thought as any single-rooted purity. If the four adequate world hypotheses have differing and somewhat complementary strengths, if they have endured so long despite their individual shortcomings, it would seem likely that a true synthesis could yield a more valuable and defensible set of metaphysical foundations for the work of the future than any pure system could hope to do. Repeatedly in the history of ideas, where competing but incompatible formulations have persisted, the outcome has not been the victory of one and rout of the other but a reorganization of the field, making it possible to transform and integrate what in untransformed guise had seemed irreconcilable. Einstein's invariance-finding principle (or algorithm) may thus help us to attain a true synthesis of world hypotheses.

In my more optimistic moments, I feel that progress is being made toward such a synthesis. The systems view of the

---

[54] Quoted in L. E. Hahn, "The Stephen C. Pepper Papers, 1903–1972," *Paunch* 53–54(1980):75.

[55] S. E. Hyman, *The Tangled Bank* (New York: Atheneum, 1962).

world, as I understand and construe it, embraces much of humanism and mechanism.[56] It shares with contextualism an emphasis on the emergence of novel properties of wholes, with formism a regard for the importance of universals and formal systems, with naturalism a conception of enduring physical entities or material systems, and with organicism a respect for the reality of ideas, values, and subjective experience.[57] The religiously inclined can as easily add on to this world view, as they have done to any other, their faith in a supernatural creator and spiritual realm.

I know that such a statement is little more than a declaration of faith and hope. Not a metaphysician by training, I doubt that I am well suited to try the synthetic task in any more serious way. Yet I am firmly convinced that metaphysics does matter, that huge benefits will accrue if satisfactory philosophical foundations can be laid for a true concert of the disciplines—the arts and humane letters alongside all the sciences. Psychoanalysis will be an indispensible component, many of Freud's ideas and discoveries finding a permanent place in any scheme to get all the resources of the human mind and heart to work together. So grounded, a new morality for human survival may take root and prosper, filling a gap that now seems to be dangerously widening.

---

[56] Cf. R. L. Ackoff, *Redesigning the Future: A Systems Approach to Societal Problems* (New York: Wiley, 1974): E. Laszlo, *Introduction to Systems Philosophy* (New York: Gordon & Branch, 1971); Paul Weiss, "The Living System: Determinism Stratified" (1969).

[57] Near the end of his life, Pepper hailed Laszlo's systems philosophy as "a breath of fresh air" toward which he felt "extremely sympathetic." He believed that it closely resembled "(and certainly includes) my own presently favored paradigm of 'selective system' for a world theory." ["On 'The Case for Systems Philosophy,'" *Metaphilosophy* 3(1972), p. 151.] Its root metaphor, the dynamic self-regulating system, he considered "possibly the most fruitful or even the correct one for a detailed synthetic comprehension of the structure of the universe." ["Systems Philosophy as a World Hypothesis," *Philosophy and Phenomenological Research* 32(1972):548.] Perhaps Pepper was right in seeing it as having a single root metaphor rather than as I have presented it as itself a synthesis of valid elements from preceding world hypotheses.

AFTERWORD

In his moving funeral oration for Freud, Ernest Jones said:

At my first meeting with him so long ago three qualities in particular produced an impression on me that only deepened as the years passed. In the first place, his nobility of character, his *Erhabenheit*. It was impossible to imagine his ever doing a petty thing or thinking a petty thought. [Then he quoted a letter from Freud to Putnam in German, in which Freud declared, "Actually I have never done a mean thing."] How many of us, if we search our hearts, could truthfully say that? Those of us who have special knowledge concerning the imperfections of mankind are sometimes depressed when we consider ourselves and our fellow men. In those moments we recall the rare spirits that transcend the smallness of life, give life its glory and show us the picture of true greatness. It is they who give life its full value. There are not many of those rare spirits and Freud was among the highest of them.[58]

Here he confesses the reason for his need to deny Freud's occasional lapses from nobility (and from the two other ideal qualities the grieving disciple went on to attribute to him—unswerving devotion to truth and to justice). Jones apparently needed to believe that certain abstract ideals could in fact be concretely achieved, as if there was no point in trying to be truthful unless one could believe that absolute, total devotion to truth was attainable. That is a prepsychoanalytic idea, not one he got from Freud. Psychoanalysis teaches, rather, that every god has clay feet, but that we can nevertheless respect and admire anyone who succeeds relatively better than others in the never-ending struggle against evil or petty impulses. The facts of Freud's life show that he was not superhuman; he was, we know, occasionally mean, untruthful, and unjust. But his own reaction to evidence of such failings in others was not to collapse in disillusion; Freud himself had no such illusions about the perfection of others, and he tried to discourage his followers from idolizing him.

[58] *The Life and Work of Sigmund Freud*, vol. 3 (New York: Basic Books, 1957), 247.

Perhaps, in that same spirit, we should not be too hard on Jones. When he wrote the words just quoted, he had just suffered a grievous loss, a blow that tends to bring up deeply infantile longings. We have enough distance, today, not to share his need to believe that Freud had somehow attained an impossible perfection. Life can have full value without the necessity of denying the existence of some imperfection in everyone.

CHAPTER EIGHT

Ethics and Excuses

The Ethical Implications of Psychoanalysis

GERALD N. IZENBERG

The question of Freud's impact on modern morality and contemporary world views can include two separate things. One is the problem of the ethical and philosophical implications of psychoanalytic theory, which requires an internal, conceptual analysis. The other is the issue of Freud's actual influence on empirical social beliefs and practices, which demands historical analysis.

Professor Holt's paper deals primarily with the former, and appropriately so. The latter is much more tenuous and difficult. There is no clear and precise way to measure the causal influence of psychoanalysis on popular thought. We can chart certain trends—in much more detail now than before—and we can speculate with more or less certainty about their meaning. We can document the spread and popularization of psychoanalysis in the 1920s through articles in magazines and the like. We can observe its consolidation in the American psychiatric establishment and in the consciousness of the educated middle class in the 1940s and 1950s and its decline in the 1960s and 1970s. We can compare its degree of relative acclimatization in Europe and the United States, noting such phenomena as its sudden rise in popularity in France, in the wake of the failure of the 1968 uprisings, after decades of marginal existence in French cultural life. But such observations cannot tell us much about the influence of psychoanalysis on the modern "decline of religion" or "crisis of morals." If indeed there have been such, psychoanalysis is as much symptom as it is cause.

At best it is possible to speculate on the social ethical impact of psychoanalysis in an impressionistic way in the light of what it seems to mean to many as an attitude to the self. Psychoanalysis definitively destroyed, after the optimism of the Enlightenment rationalist tradition, any presumption of innocence in the human being generally and in the child particularly. It restored an older religious image in secular form. It exposed us to ourselves as lustful, aggressive, needy, murderously enraged, and as deceptive about all of these. By the same token, it made us suspicious of ourselves and of one another, inculcating an attitude of skepticism and watchfulness reminiscent of the original Puritan communities. A. A. Brill, one of the early disciples of Freud, describes how they would all carefully watch one another in the quest for hidden motives, pouncing upon the slightest deviation from routine behavior; one needed only to pick up a fork at dinner with the wrong hand to be met with the inquisitorial "Why did you do that?" We can no longer naively take our behavior at face value.

There is, of course, a more benign way of putting this. We have been made introspective, self-aware, psychologically minded. But it is more in keeping with the moral thrust of Freud himself to stress the accusatorial. That thrust was in the direction of demystification, which can be defined crudely as exposing the low-mindedness behind high-mindedness but perhaps more crucially as preventing self-deception. The analytic attitude contains implicitly an ethic of not being in bad faith—an ethic of honesty but with stress on the negative, on not being *dishonest*—because of the inbuilt awareness that our natural tendency *is* to self-deception.

Paradoxically, psychoanalysis provided us at the same time with an "attitude of excuse," a saving way of looking at our "badness," to help us cope with its radical accusations. I do not mean by this the reification that existentialists, among others, charge is built into Freud's structural psychology and that subtly enables us to evade responsibility through such locutions as "My id did it." Freud intended no way out here. Nor do I mean Freud's rigid psychic determinism on the philosophical level, with its negative implications for free will and responsibility. Freud was inconsistent on this in any case, and his underlying metaphysics of science was not the influential aspect of psychoanalysis.

By "attitude of excuse," I mean a number of things. For one, Freud's biological foundation and developmental perspective meant that in our impulses and psychic conflicts, no matter how "evil" by conventional moral standards, we could recognize nature and not have to fear inhuman monstrousness. The sexual perversions, for example, stamped us not as bizarre deviants but as fixated somewhere early in a normal developmental process. "We are not," said Freud, "as rational as we thought we were, but we are not as unnatural as our irrationality makes us seem." To recognize the child in us at the heart of neurotic behavior is relatively comforting, when we feared much worse. Moreover, the crucial events generating neuroses also occurred in childhood, when, as Freud once tried morally to console one of his patients, we naturally have less control over impulses, fears, and anxieties.

In the second place, neurosis within the psychoanalytic framework is seen as illness. The medical model has quite specific moral implications; it licenses a lessening of responsibility for questionable behavior and gives the neurotic the status of patient with a right to obtain help. For the analysand to see himself as in some sense ill may also be therapeutically necessary, for example, to counteract those powerful resistances that are manifested in the form of guilt; some analysts will even explicitly argue the definition of neurosis as illness in the course of analysis, so as to cope with the patient's self-condemnation. The medical model undoubtedly creates severe legal as well as philosophical problems; for psychoanalysis holds, theoretically, that so-called accidental or irresponsible behavior is unconsciously *willed* and therefore in some sense controllable. That is, the very possibility of cure in psychoanalysis depends on the ability of the neurotic to recognize and assimilate insight and change his behavior. But the ambiguity of psychoanalysis on this point has functioned as a means to hold moral judgment of conduct in abeyance while the therapeutic efforts are made to return that conduct to the full responsibility of the neurotic.

No doubt there have been other "popular" effects of psychoanalysis. It has perhaps fostered a degree of permissiveness toward previously restricted behavior, at least to the extent that it sanctions legitimate, socially nonthreatening expressions of the instincts, and Freud's attack on religion as the perpetuation of infantile dependence is fairly widely known. But the first of these ought not to be exaggerated, despite the occasional complaints of conservatives against modern "permissive" doctrines of child rearing. Freud's permissiveness was carefully restricted to the consulting room and freedom of word and idea association. There *has* been a tradition of "wild" psychoanalysis, with the occasional preaching of the glories of "polymorphous perversity." But political and social radicals have more often seen psychoanalysis as socially conformist rather than radical in its implications. Psychoanalysis, from their perspective, claims that problems are internally rather that socially caused and preaches adaptation to an exploitative reality.

Let me now turn to the internal analysis of the moral implications of psychoanalytic theory. I would agree that Freud personally, and psychoanalysis theoretically, affirmed traditional moral standards with the exceptions that Holt notes: Freud did stand for a freer sexual ethic than the traditional Victorian one, and he did wish to lessen the condemnation of the superego for wishes and thought as opposed to deeds. However, I question Holt's statement that Freud believed that analyzed people would be more ethical than others. Freud is, in fact, on record as saying exactly the opposite. "Why," he once wrote, "should analyzed people be altogether better than others? Analysis makes for unity, but not necessarily for goodness." In this sense, an important moral shift is represented in the therapeutic thrust of psychoanalysis. It puts a premium on the integrity of personality through consciousness—that is, on conscious control and choice as opposed to unconscious drivenness. In this sense, there is a parallel with the existential ethic of responsibility. What is of importance to both is not so much the content of what one does but the ability to avow consciously, and to affirm responsibility for, *whatever* one does. Since the focus of both thought systems, each in its different way, is on self-deception through compartmentalization and concealment of intentions, the aim of both is a self-demysti-fication leading to transparency and self-awareness.

Moreover, I do not think that the moral standpoint of psychoanalysis is adequately characterized by Piaget's and Kohlberg's categories of moral development. The stage of "postconventional" morality represents the ability of the moral agent to be flexible in the application of moral absolutes, and this supposedly corresponds to the situation of the analyzed patient whose superego is no longer rigid, incapable of mod-ifying to demands in the light of reality or multiple moral imperatives. There is some justice in this comparison. In an important sense, however, psychoanalytic theory undercuts the very idea of moral absolutes or a transcendental ethic. The psychoanalytic explanation of the origin of morality operates a double reduction: the genetic explanation reduces morality to the internalized parents, and the biological grounding of psychoanalysis gives prohibitions an essentially biological–social utilitarian function. They operate in the interests of the reality principle, not in the interests of transcendent moral standards or a categorical imperative. The reality principle is in fact the principle of rational or enlightened self-interest, a prudent version of the pleasure principle.

This is not to say that Freud was a simple utilitarian. It is certainly true that Freud unproblematically believed that the pleasure principle supplied the purpose of life. As Philip Rieff has pointed out, however, Freud was not a simple apostle of pleasure. Pleasure was, in the energy metapsychology of his theory, a negative idea—a decrease through discharge of unpleasurable tension. Since tension was never reducible to zero (which would mean death) and was in fact, repeatedly built up again, pleasure was never wholly or finally achievable.

Beyond this, and more importantly, pleasure was, for Freud, despite the biological base of his theory, ultimately a moral and not a biological category. Libidinal behavior meant authentic selfhood, self-motivated as opposed to alienated behavior. This emerged in a number of ways. The category of sublimation was one; the idea of behavior that satisfied sexual instincts without being sexual made sense only if what joined the two was the idea that both expressed the natural impulses of the self. But Freud was even more directly revealing on this point. In the fight against C. G. Jung's reduction of all drives to one basic instinct, Freud made an argument for preservation of a dual instinct theory, libido and ego, that revealed the true nature of the duality at the heart of instinctual conflict. In his words, "The individual does actually carry on a double existence: *one designed to serve his own purposes,* and another as a link in a chain, in which he serves against, or at any rate without, any volition of his own" (emphasis added). Examined closely, this argument proves to be not one in favor of two instincts but rather an argument for considering the sexual instinct in two modes—as it serves the agent's own purposes or as it does not.

For Freud, both modes were necessary, one for gratification, the other one for survival (whether of self or species). But when the individual subordinates his instinctual desire to the needs of society and the species, he must, in the analytic ethic, do it knowingly, consciously recognizing the human need for suppression or sublimation. Hence Freud believed, for example, that the incest taboo had to be enforced on the human race, but *not* as a taboo—that is, not automatically, categorically, without self-awareness of the desire; otherwise true freedom, in the psychoanalytic sense, was impossible. "Anybody," he once wrote, "who is to be really free and happy in love must have surmounted his respect for women and have come to terms with the idea of incest with mother and sister."

Abandoning a taboo morality meant overcoming the alienation of absolute moral codes: it meant recognizing the purposes of the self and of humankind in the fiats of morality.

This brings me to a second major point made by Holt, the issue of the narcissistic thrust of psychoanalysis as a whole. In the context of his times, Freud stood for the reaffirmation of the ideals of liberalism and romanticism. I have argued elsewhere that, through his theoretical categories (primarily instinct theory), Freud actually reaffirmed his belief in original human freedom despite his empirical clinical discoveries of unfreedom, irrationality, and infantilism. Freud struggled to retain the individualist and rationalist perspectives in the face of irrationality.

But it is also true that, from a contemporary point of view, the Freudian stress on self-realization, even in the measured form of the reality principle, looks like a contribution to narcissism. This was not Freud's meaning or intention. His ideal was culture and sublimation, not personal instinctual gratification. Specifically in relationship to the problem of narcissism, it was Freud who warned that we must love others in order not to fall psychologically ill. Nevertheless, psychoanalysis is part of the rationalist individualist tradition, and that tradition stressed the asocial notion of freedom—freedom defined in negative terms: individual and society are seen within that tradition as opposed forces, since the satisfaction of each comes at the expense of the other.

There is another aspect to this problem. As Holt points out, psychoanalysis, with its resolutely individualist standpoint, has no sociology. This means that the focus of moral responsibility, and blame, is totally on the individual. In the case of narcissistic behavior, for example, it is seen as the result of a purely internal need or self-concept—the grandiose self. Ultimately there is no *theoretical* room in psychoanalysis (whatever *ad hoc* clinical concessions are made) for any awareness of the impact of the social environment on drives and needs. Otto Kernberg, one of the major contemporary theoreticians on the problem of narcissistic personality disorders, argues that the crucial events behind such disorders occur intrapsychically in the very young child before the impact of society is felt. He, like all modern analysts, weighs the impact of the early nurturing figures, but he misses the seemingly obvious basic fact that parents are social agents. The modes of parenting—absence, demands, conditional loving, and so on—are not only reflections of the early intrapsychic structures of parents but also manifestations of social ambitions and anxieties produced by social structures and social goals. This relentless individualism of perspective illustrated by Kernberg is one of the important reasons that psychoanalysis was found acceptable in the United States. It was in this respect in keeping with the American ideology of the individual's total responsibility for his own success or failure.

So far as an ethic of health is concerned, there is, we can perhaps say, nothing to add to Freud's famous formula that happiness in life is the ability to love and to work. The problem is what prevents that from being possible. And here we can agree with Holt that the purely intrapsychic approach prevents adequate social analyses. The latter cannot simply be superadded to psychoanalysis. It requires a wholesale revamping of psychoanalytic metapsychology, beginning with its solipsistic biological and energy theories.

CHAPTER NINE

Freud's Influence on the Moral Aspects of the Physician–Patient Relationship

EUGENE B. BRODY

My comments stem mainly from my identity as a physician, a person to whom society grants access, for health-related purposes, to the minds and bodies of others—sometimes, even, unwilling others. Privilege of access carries moral responsibility. The primary and traditional medical responsibility is not to harm. Is the physician also obliged to help the patient become a "better" person? Most would answer in the negative, although their expectations of patient compliance suggest that this may well be so. In fact, such an expectation would fit the socially reinforced paternalism of the physician. The answer for the psychoanalyst, however, must always be yes, if increased self-knowledge is regarded as intrinsically "good" or morally desirable. Self-knowledge, after all, is granted highest value as the goal of the clinical psychoanalytic process. It signifies, in Freudian terms, freedom from the bondage of the unconscious and, therefore, increased conscious responsibility for one's own behavior. Freud's legacy, I believe, includes the moral obligation upon the physician (exemplified by the analyst) to make his patient a "better" (i.e., a more moral) person

by increasing his knowledge of himself and, thereby, his capacity to behave in a more authentic and responsible manner.

On the other hand, some doubt that such knowledge allows or requires one to behave more responsibly. Further, responsibility for one's behavior in the sense of claiming it truly
as one's own does not necessarily mean that one will act more
ethically. The nonanalyst may even ask whether psychoanalysis can, in fact, truly increase self-knowledge. Opinions about
its capacity to do so vary with acceptance (1) of the theory's
view of human nature or the human self and (2) of the validity
of psychoanalytic knowledge. Judith Tormey and I have recently argued that specific features of the clinical psychoanalytic process involve, as inevitable consequences, epistemological dilemmas and uncertainties that influence what is
regarded as psychoanalytic knowledge and make it impossible
to apply any single criterion of truth to it.[1] We argue, also, that
the patient's subjective reality as it emerges during the analytic
process does not exist independently. It is, rather, a consequence of the analyst's interventions,—an intersubjective reflection of continuing reciprocity between the two participants.
The object of knowledge for both is the unit, patient–analyst.
This means that the knowledge regarded as illuminating the
patient's true "self," or as representing what was previously
repressed and determining behavior without awareness, may
in fact not be that at all. It may represent, rather, a symbolic
or rational construct created jointly by analyst and patient, and
thus with a base in their shared reality, but limited by culture,
language, and analytic theory. This last is especially important.
As Tormey and I have pointed out, the interpretations that
evoke new material seen as indicating the repressed simultaneously constitute particularistic hypotheses and, through
the study of the responsive associations to them, ways of testing general statements: "the theory is tested by the use of
measuring instruments which it, itself, has created." Beyond

[1] Eugene B. Brody and Judith F. Tormey, "Clinical Psychoanalytic Knowledge—An Epistemological Inquiry," *Perspectives in Biology and Medicine*
24 (Autumn 1980): 143–59.

this, the analyst, chief architect of the knowledge based on the study of the analyst–patient unity, is already limited as a knowing instrument by the theory within which he operates. The self-knowledge, then, pursued by patient and analyst seems more likely to meet a criterion of coherence rather than of correspondence to something "actually there."

Is the obligatory search for self-knowledge the same as the search for intelligibility? The last would meet, at least, a criterion of coherence. The relevant Freudian concept in the public domain that bears directly upon contemporary views of personal freedom and responsibility is that of the unconscious. Some variant of the freedom–responsibility concept is contained within every group's socially inherited rules for living. To the best of my knowledge there has been no culture in which individual freedom has not been conceived as limited by attributing responsibility for personal action on occasion to some force out of awareness, or beyond the subject's conscious control.

Freud's theory, as I suggested earlier, has something in common with other cultural attributions of unwished for impulses and behavior to dark recesses within our own souls. The idea of a mind that can be divided against itself long antedated Freud. Reconciling that division, however, required the citizen or a healer-prophet to supplicate a supernatural presence and then interpret its cryptic utterances. It was not until Freud that an active, generally available intervention was devised and placed in the professional healer's armamentarium. Freud's technique—which can be acquired by any introspective, educated, and motivated person—allowed supplication to be replaced by therapy. Coming at a time when psychiatry was gaining acceptance as part of medicine, it was part of the long, slow historical secularization of concerns with the mind, especially its "good" and "evil" aspects. Psychoanalysis was based, though, on the same assumption made by the supplicators: *all* behavior and *all* utterances, however unclear, are ultimately intelligible. That is, utterances and acts (and feelings) have a certain logical coherence and correspond systematically to factors that can be identified. In *particular* cases—in which motivated people present themselves for treatment—the analyst becomes specifically responsible for discovering the intelligibility in their behavior. Then he must make it accessible to them, facilitating their capacity to reflect upon themselves and thus eventually to change. Freud did not, though, assume responsibility for the direction of change. His working hypothesis seemed to be that conscious awareness of previously unconscious factors was the analytic goal and that decision making, on the basis of the newly accessible information, remained the patient's own responsibility. Writing of "virtually immortal" id impulses, he stated: "They . . . can only lose their importance . . . when they have been made conscious by the work of analysis."[2] His position was clearly stated in 1923 when he described the therapeutic aim of psychoanalysis as "not . . . to make pathological reactions impossible, but to give the patient's ego freedom to decide one way or the other."[3] This freedom to make significant choices on the basis of rational factors available for reflection does not imply that the choices would necessarily accord with accepted moral values. It suggests only a certain authenticity of the knowledge informing them. They would not pursue the symbolic or disguised aims of repressed and, therefore, unconscious intentions. As intending would no longer have meaning of which the intender is not aware, the division between known and unknown elements in oneself would be replaced by a more integrated, unified conscious self as the basis for action.

Is there a general moral obligation to understand oneself inherent in Freud's assumption of the general intelligibility of behavior? Does Freud's theory, in sum, impute moral value to being authentic, deciding and acting without the distortion of hidden wishes and conflicts? The implied goal may be less to uncover repressed, possibly evil, influences on behavior than

[2] Sigmund Freud, *New Introductory Lectures, Standard Edition*, vol. 22 (London: Hogarth, 1964), p. 74.
[3] Sigmund Freud, *The Ego and the Id, Standard Edition,* vol. 19, p. 50n.

to be coherent. Hypothetically, this could be to be "coherently evil."

Is the moral obligation more apt to reside in the therapeutic goal of psychoanalysis, which is "freedom to decide"? I think that Freud personally felt this obligation when he recognized that the concept of an unconscious permits disavowal of personal responsibility for behavior regarded as derivative of repressed, unacceptable intentions. This sense of obligation is contained within his statement quoted by Holt: "I shall perhaps learn that what I am disavowing not only 'is' in me, but sometimes 'acts' from out of me as well."[4]

A contemporary psychoanalyst, the late Lawrence Kubie, alluded, in 1952, in a somewhat similar vein, to "the symbolic process of self-expression in . . . language and action," in contrast to "the symbolic process of self-deception,"[5] by which he meant hiding one's deeper goals, memories, or conflicts from oneself. Kubie's emphasis on the importance of freedom from unconscious determinants of behavior and what he called the "self-deceiving process" also suggest a personal feeling of obligation to seek that freedom.

Disavowal *without* the intentional connotation of self-deception is suggested by a statement of another contemporary analyst, the late George Klein. "Repression" he wrote, maintains a "gap in comprehension." It is "the refusal to acknowledge the meaning of a tendency, but not the tendency, itself; the gratifying aspect of the tendency has not been renounced . . . meanings are 'lived out' without comprehension."[6] The gap in comprehension, here, is not identified as a conscious state which is experienced and can be dealt with as such. The primary experience is of gratification or, perhaps, aversion if another suggests a failure of personal responsibility or com-

[4] Ibid., p. 133.
[5] Lawrence S. Kubie, "Problems and Techniques of Psychoanalytic Validation and Progress." In *Psychoanalysis as Science*, ed. E. Pumpian-Mindlin (Stanford, Calif.: Stanford University Press, 1952), 46.
[6] George S. Klein, *Psychoanalytic Theory: An Exploration of Essentials* (New York: International Universities Press, 1976), 241–42.

prehension. The examining analyst, though, may make it possible for the patient to understand the gap as reflecting inner
incoherence or conflict. A sense of inner conflict or indefinable
dissatisfaction associated with a failure of sensed integrity does,
in fact, motivate some people to seek psychoanalytic assistance. The idea of self-knowledge, beyond consciousness and
awareness of being conscious, conveys a suggestion of responsibility—the moral awareness of the intending self and
its possibilities as its own agent. Perhaps the most explicit
statement of this fitting the current cultural emphasis on the
self is that of Roy Schafer. He conceives of psychoanalytic
change in terms of intending and acting so as to make new
arrangements. The patient lives increasingly in terms of the
self as agent—doing the things from which he or she was
previously suffering.

My own clinical experience suggests that self-knowledge
may produce increasing degrees of accepted responsibility (and
intentional freedom) as the possibilities for attributing behavior
to obligatory sources is reduced. A famous, accomplished man,
for example, who had come to therapy complaining of depression and fatigue, rejected a professional suggestion from a
junior colleague without considering its obvious merits. He
explained himself by saying that the suggestion was useless
and possibly hazardous. It was not difficult for him to become
aware of his sense of the colleague as competitive and unreasonably ambitious. Analysis of the situation, then, proceeded
from identifying his repressed fear of being overthrown by a
symbolic son to a series of earlier life determinants of pre-
conscious feelings of inferiority, jealousy, envy, and rage involving his relations with his parents (he was the only child
of a young mother and an elderly, ferociously achieving father).
With each new development he was able more clearly to grasp
the contemporary reality of the situation and eventually to
reconsider the younger man's ideas—which, in the long run,
turned out to be beneficial for both.

In this instance progressive degrees of self-knowledge
(freedom from unconscious factors) may be regarded as af-

fording increasing freedom in decision making and a more authentic claiming of responsibility for behavior in regard to the colleague. The patient, of course, had never disavowed his act directly. He had disavowed it indirectly by a false explanation (i.e., the suggestion was no good). In this sense he was self-deceiving. Self-knowledge made the false explanation untenable and, thus, forced him into more responsible behavior (i.e., identifying more valid "reasons for" his behavior).

The problem, of course, with drawing conclusions from psychoanalytic patients is that their coming to treatment means that, like Freud, they have already assumed responsibility for what is unknown in themselves. They have embraced the obligation to seek intelligibility in their own seemingly unintelligible acts of feelings. They may not comprehend the meaning of behavior in which they continue to engage because it is gratifying. They may even deplore it. But they do not attribute it to an independently operating unconscious or, as primitives or the early Greeks might have, to an external force. It seems to me that it is not so much the awareness of inner division or (abstractly) of a gap in comprehension as that coupled with the obligation to clarify the obscure which reflects the social impact of Freudian views. This allows—within limits—the cultural acceptance of repeated fruitless or destructive acts, exploitative gratification, or disturbing ideas and feelings without moral condemnation as long as they are regarded clinically or called symptomatic. Moral redemption requires the person's willingness to work therapeutically toward self-awareness and self-knowledge. The validity of this self-knowledge is not at issue.

Clinically achievable "self-knowledge" as it is currently conceived appears to me to have much in common with Plato's *epistēmē*, the truth opposed to ignorance, opinion, and—most importantly—illusion. It is a truth essential to the health and integrity of *psùche*—psyche, soul, mind, or self. That truth today is invoked in support of rights considered essential to human well-being. Human rights (to freedom, autonomy, justice, opportunity) are justified in part on epistemological grounds.

We are so convinced of the value of knowledge gained by looking into and reflecting upon ourselves that we consider it as self-evidently constituting a claim upon others to respect us as individuals.

There is, though, an apparent paradox here. The capacity for self-knowledge is linked at the same time to the possibility of increased personal freedom and decreased social freedom. If we regard all human behavior as potentially intelligible and all people with intact brains as capable, with guidance, of reflecting upon their own behavior and assuming responsibility for change, we move toward what Nicholas Kittrie called the "therapeutic state." Behavior transgressing public moral standards and, therefore, labeled as wrong or criminal can be regarded as "symptomatic" and, by the same token, as an indication for "therapy." It is considered to reflect a hidden (i.e., potential) intelligibility of which the actor is himself unaware. The assumption of potential intelligibility implies potentially identifiable but presently unknown ("unconscious") causes for the behavior in question and hence that it was not freely willed or intended. Freedom of will, here, does not refer to an unlimited range of options or potential decisions. It refers, rather, to being aware of at least the immediate range of available options and the factors that might influence decisions about them. An adolescent patient's decision about a vocational choice, for example, might be made on the (unidentified) basis of unconscious anger against the father but rationalized by a variety of plausible reasons. When, through psychotherapeutic work, the patient becomes aware of the anger, it may still be a reason for not choosing the same vocation as the father; but in this instance we regard the decision as "freely" in contrast to "symptomatically" or blindly made. In this instance Freud's views of unconscious motivation suggest a moral obligation by the decision maker to know and hence to act freely and rationally. It compels the obligation on the part of the physician to understand the patient's capacity for knowing and if he— the physician—chooses not to move the patient toward self-knowledge, to have good reasons for taking some other therapeutic course.

We assume that if the intelligibility behind his immoral acts is discovered and reflected on by the actor, this offers the possibility of change. The change need not, however, fit current standards of what is right, preferred, or desirable. Holt noted that it is "a real empirical question whether psychoanalyzed people do restrain themselves from immoral acts as well as those who succeed in continuously repressing the corresponding wishes." So the act of help seeking, even though it eventuates in his increased self-knowledge, may have greater public significance as expiation than private significance as a responsible change in life-style. The therapist, furthermore—in the Freudian tradition—is the patient's agent, not explicitly concerned with his morality or conformity except insofar as they provoke destructive retaliation from others or interfere, through increased internal tension, with his own functioning. If he follows Freud, his therapeutic goal is not to preclude pathological (in this instance immoral or criminal) acts but to give the patient "freedom to decide."

The implication of a *required* search for insight nonetheless confronts us in a new guise with the old prescription. The old prescription was that the deviant (who might have been considered wicked, a witch, or possessed) should be subjected through psychological or physical manipulation to an intense affective experience which—if he survived it—would prove him to be as good or make him a better person. A newer way out of the semanticized moral dilemma may lie in B. F. Skinner's operant conditioning. Therapy based on this idea carefully avoids the notion of intelligibility or that the "person" himself is involved. The change is limited to ways of behaving or acting and does not imply anything about what goes on in the mind, designated conceptually as a "black box." The moral self as a responsible, intending, self-aware agent becomes irrelevant.

Is it moral for the physician to manipulate the patient in the interest of any agency other than the patient's own self? This, of course, is already routinely done. Quarantine and case reporting for certain communicable diseases or for disturbances such as epilepsy may protect the community while

placing the patient at a personal disadvantage. Freud's contribution to our understanding of this problem bears on the
influence of unconscious factors in clinical and policy situations. This seems to me intertwined with the influence of these
factors on what we regard as knowledge. To the degree that
the clinician is not responding to the reality of his patient but
to his own historically rooted and unconsciously based needs
and perceptions (i.e., in terms of his own transference); he
may make interpretations, prescribe actions, or otherwise engage in behavior not useful to his patient. At worst he may
exploit his patient, however unwittingly, to gratify his own
needs, which may be to serve the institution that employs him.
At the same time, he becomes an even more serious intruder
into what can be considered the "objective" facts of the case
than the clinician who suffers only from his theoretical
preconceptions.

The therapist's knowledge, on the basis of which his decisions are regarded as logically necessary, may in fact com-
prise systematic distortions consequent to his own transference processes. This is a special instance of the general problem
faced by empiricists who require clear rules of correspondence
between observed phenomena on one hand and the ideas or
theories which presumably flow from them on the other. The
analytic theory in particular needs independent validating from
outside the analytic situation. I think this—and here I part
company from Holt—because I believe that the training analysis poses a serious epistemological dilemma. Holt regards the
training analysis as helping "the neophyte learn his own characteristic ways of distorting reality," so that he may compen-
sate for them. He regards this as approximating "a set of transformations that yield an invariant lawfulness and the possibility
of true knowledge." Others regard the training analysis as
removing problems within the student analyst, with the result
that he will miss important interpretations. I can agree that
as the analyst works *within his theory* or within his metaphor,
the training analysis contributes to his capacity to observe his
own discrepant feelings (e.g., love or hate for his patient) or

discrepant behaviors (e.g., dreaming about his patient). The possibility of "true knowledge," though, it seems to me, is even more remote in consequence of the training analysis. How can observations be used to test a theory when they are made by a cognitive faculty already shaped by that theory?

This brings me, finally, to Holt's statements that "psychoanalysis lacks a social vision in which all members of a society are equally important" and that it "shares with liberalism the short-sightedness of giving overriding importance to personal freedom."

An adequate discussion of these issues would include Rousseau and the social contract, Mill and the variants of utilitarianism, and the entire field of social ethics. I will limit myself, therefore, to drawing an analogy between issues I have faced as a policy person and those which arise in the psychoanalytic situation. As an adviser on population policy in a small, mainly agrarian, largely illiterate country, for example, it was clear to me that, regardless of specific recommendations, the interests of those most remote from the decision makers (i.e., the most disadvantaged members of society) would be submerged.[7] I also felt it unethical to manipulate the motives of the largely illiterate public in a manner designed to preserve the illusion of free choice. (These views suggest my agreement with Holt's observation of Freud's lack of social vision and my disagreement with his view of Freud's overemphasis on personal freedom—although I am not yet sure of the meaning of personal versus social freedom.) In this situation, at any event, I appreciated John Rawls's general idea that policy should be made behind a "veil of ignorance" (i.e., by persons who do not know where they will be in the social structure influenced by the policy). "Liberty and opportunity, income and wealth, and the bases of self-respect" should be distributed, according to Rawls, so as to maximize the minimum share. This emphasized for me, consulting in a society dominated by a small,

[7] Eugene B. Brody, "Reproductive Freedom, Coercion and Justice: Some Ethical Aspects of Population Policy and Practice," *Social Science and Medicine*, Vol. 10 (New York: Pergamon Press, 1976), 553–557.

educated elite, the value of "a social contract insuring a favorable outcome and an increased sense of personal worth for its least advantaged members," as well as one that emphasized "the primary value of self-esteem in a just society, based on an equitable and publicly visible distribution of rights and liberties."[8]

It seems likely to me that policymakers' attitudes toward the people in regard to whom they have an authoritative position and for whom, with minimal direct consultation, they create contexts must be strongly subject to unconscious factors—perhaps in some ways analogous to the analyst's countertransference. One of these factors possibly compatible with conscious desires I take to be a wish for stability. This is paradoxical both for the analyst, as he is dedicated to personal change in his patient, and for the policymaker, as he is dedicated to constructive social change. Both, however, require maintenance of significant elements of the status quo in order to continue to do their own work and to preserve their own roles, vis-à-vis the patient or vis-à-vis society. For the analyst it may contribute to recommending continuing analysis for his patient and hence continuing dependence, when a trial of life on one's own is needed. The clinical process aimed at freedom to choose and at autonomy may, without recognition by either party to the transference–countertransference relationship, contribute instead to clinically rationalized self-doubting and dependency.

[8] Ibid., pp. 553–54.

Index